A Natural History of Human Morality

A Natural History of Human Morality

Michael Tomasello

Harvard University Press

Cambridge, Massachusetts
London, England
2016

Printed in the United States of America

First printing

Library of Congress Cataloging-in-Publication Data

Tomasello, Michael.
A natural history of human morality / Michael Tomasello.
pages cm
Includes bibliographical references and index.
ISBN 978-0-674-08864-1 (hardcover : alk. paper) 1. Ethics—History.
2. Ethics, Evolutionary. 3. Philosophical anthropology. I. Title.
BJ1298.T66 2015
170.9—dc23

2015010960

For Chiara, Leo, Anya, *and* Rita

Contents

Preface ix

1.

The Interdependence Hypothesis 1

2.

Evolution of Cooperation 9

Foundations of Cooperation 10

Great Ape Cooperation 20

Kin- and Friend-Based Prosociality 34

3.

Second-Personal Morality 39

Collaboration and Helping 42

Joint Intentionality 50

Second-Personal Agency 57

Joint Commitment 64

The Original "Ought" 78

4.

"Objective" Morality 85

Culture and Loyalty 88

Collective Intentionality 92

Cultural Agency 97

Moral Self-Governance 107

The Original Right and Wrong 121

Coda: After the Garden of Eden 129

5.

Human Morality as Cooperation-Plus 135

Theories of the Evolution of Morality 137

Shared Intentionality and Morality 143

The Role of Ontogeny 154

Conclusion 158

Notes 165

References 169

Index 187

Preface

A Natural History of Human Morality is a companion volume to my 2014 book *A Natural History of Human Thinking*. The parallel titles are appropriate because in both volumes I propose the same two-step sequence in the evolution of human social life: first, new forms of collaborative activity, and then new forms of cultural organization. In the first volume I attempted to spell out the species-unique kinds of thinking that emerged from these new forms of social life. In the current volume I attempt to explicate how these new forms of social life structured the way that early humans came to engage in moral acts that either subordinated or treated as equal their own interests and the interests of others, even feeling a sense of obligation to do so. This moral attitude or stance did not—and still does not—win out consistently in individuals' actual decision making, of course, but it does make those decisions, whatever their outcome, moral decisions.

I have been gathering my thoughts for this volume over the past five years or so, beginning with a seminar on the evolution of human cooperation held in the fall of 2009 here at the Max Planck Institute for Evolutionary Anthropology, and continuing with a similar seminar on the evolution of human morality held in the winter of 2012–2013. Many interesting and fruitful discussions in those seminars have shaped my thinking on these issues significantly, and I thank all of those who participated. I also had a number of very useful discussions during this same time period with Sebastian Rödl, who helped me with some difficult philosophical concepts.

In addition, a number of people read and provided very useful commentary on earlier versions of the manuscript. In particular, one or another draft version was read by Ivan Cabrera, Robert Hepach, Patricia Kanngiesser, Christian Kietzmann, Berislav Marusic, Cathal O'Madagain, and Marco Schmidt. I thank them all for their extremely helpful comments and suggestions. I would like to single out for special thanks Neil Roughley and Jan Engelmann, who engaged with me and the manuscript especially deeply and on multiple occasions. For certain, the manuscript is much more coherent for all of their insights. I also thank Andrew Kinney, Richard Joyce, and an anonymous reviewer from Harvard University Press for their comments on the manuscript as well.

Finally, as with the first volume, my deepest gratitude goes to Rita Svetlova, with whom I have discussed extensively all of the most important ideas in this volume—and others that she helped me to let go of—to the great benefit of the final product. I dedicate this book to her and our children.

A Natural History of Human Morality

I

The Interdependence Hypothesis

The commitments that bind us to the social body are obligatory only because they are mutual; and their nature is such that in fulfilling them one cannot work for others without at the same time working for oneself.

—JEAN JACQUES ROUSSEAU, *THE SOCIAL CONTRACT*

Cooperation appears in nature in two basic forms: altruistic helping, in which one individual sacrifices for the benefit of another, and mutualistic collaboration, in which all interacting parties benefit in some way. The uniquely human version of cooperation known as morality appears in nature in two analogous forms. On the one hand, one individual may sacrifice to help another based on such self-immolating motives as compassion, concern, and benevolence. On the other hand, interacting individuals may seek a way for all to benefit in a more balanced manner based on such impartial motives as fairness, equity, and justice. Many classical accounts in moral philosophy capture this difference by contrasting a motive for beneficence (the good) with a motive for justice (the right), and many modern accounts capture the difference by contrasting a morality of sympathy with a morality of fairness.

The morality of sympathy is most basic, as concern for the well-being of others is the sine qua non of all things moral. The evolutionary source of sympathetic concern is almost certainly parental care of offspring based in kin selection. In mammals this means everything from providing sustenance to one's offspring through nursing—regulated by the mammalian "love hormone" oxytocin—to protecting one's offspring from predators and other dangers. In this sense basically all mammals show sympathetic concern, at the very least for offspring, but in some species for selected nonkin as well. In general, the expression of sympathy is relatively straightforward. There may be some cognitive complexity in determining what is good for one's offspring or others, but once that is determined, helping is helping, with the only serious conflict being whether the sympathy that motivates the helping act is strong enough to overcome any self-serving motives involved. Acts of helping motivated by sympathetic concern are altruistic acts freely performed and are not accompanied, in their purest form, by a sense of obligation.

In contrast, the morality of fairness is neither so basic nor so straightforward—and it may very well be confined to the human species. The fundamental problem is that in situations requiring fairness there is typically a complex interaction of the cooperative and competitive motives of multiple individuals. Attempting to be fair means trying to achieve some kind of balance among all of these, and there are typically many possible ways of doing this based on many different criteria. Humans thus enter into such complex situations prepared to invoke moral judgments about the "deservingness" of the individuals involved, including the self, but they are at the same time armed with more punitive moral attitudes such as resentment or indignation against unfair others. In addition, they have still other moral attitudes that are not exactly punitive but nevertheless stern, in which they seek to hold interactive partners accountable for their actions by invoking interpersonal judgments of responsibility, obligation, commitment, trust, respect, duty, blame, and guilt. The morality of fairness is thus much more complicated than the morality of sympathy. Moreover, and perhaps not unrelated, its judgments typically carry with them some sense of responsibility or obligation: it is not just that I want to be fair to all concerned, but that one *ought* to be fair to all concerned. In general, we may say that whereas sympathy is pure cooperation, fairness is a kind of cooperativization of competition in which individuals seek balanced solutions to the many and conflicting demands of multiple participants' various motives.

Our goal in this book is to provide an evolutionary account of the emergence of human morality, in terms of both sympathy and fairness. We proceed from the assumption that human morality is a form of cooperation, specifically, the form that has emerged as humans have adapted to new and species-unique forms of social interaction and organization. Because *Homo sapiens* is an ultracooperative primate, and presumably the only moral one, we further assume that human morality comprises the key set of species-unique proximate mechanisms—psychological processes of cognition, social interaction, and self-regulation—that enable human individuals to survive and thrive in their especially cooperative social arrangements. Given these assumptions, our attempt in this book is (1) to specify in as much detail as possible, based mainly on experimental research, how the cooperation of humans differs from that of their nearest primate relatives; and (2) to construct a plausible evolutionary scenario for how such uniquely human cooperation gave rise to human morality.

The starting point is nonhuman primates, especially humans' nearest living relatives, the great apes. As in all social species, great ape individuals living in the same social group depend on one another for survival—they are interdependent (Roberts, 2005)—and so it makes sense for them to help and care for one another. Moreover, as in many primate species, great ape individuals form long-term prosocial relationships with specific other individuals in their group. In some cases these relationships are with kin, but in other cases they are with unrelated groupmates, or "friends" (Seyfarth and Cheney, 2012). Individuals depend on these special relationships to enhance their fitness, and so they invest in them, for example, by preferentially grooming their friends or supporting them in fights. The evolutionary starting point for our natural history of human morality, therefore, is the prosocial behavior that great apes in general show for those with whom they are interdependent, namely, kin and friends.

Tomasello et al. (2012) provide an account of the evolution of uniquely human cooperation that focuses on how, from this great ape starting point, early human individuals became ever more interdependent with one another for cooperative support. The interdependence hypothesis—whose basic framework we adopt here—is that this took place in two key steps, both of which involved new ecological circumstances that forced early humans into new modes of social interaction and organization: first collaboration and then culture. The individuals who did best in these new social circumstances were those who recognized their interdependencies with others and acted accordingly, a kind of cooperative rationality. Although the individuals of many animal species are interdependent in various ways, early humans' interdependencies thus rested on a new and unique set of proximate psychological mechanisms. These new and unique mechanisms enabled individuals to create with others a plural-agent "we," as in what "we" must do to capture a prey or how "we" should defend our group from other groups. The central claim of the current account is that the skills and motivation to construct with others an interdependent, plural-agent "we"—that is, the skills and motivation to participate with others in acts of *shared intentionality* (Bratman, 1992, 2014; Gilbert, 1990, 2014)—are what propelled the human species from strategic cooperation to genuine morality.

The first key step occurred hundreds of thousands of years ago, as a change in ecology forced early humans to forage together with a partner or else starve. This new form of interdependence meant that early humans now extended their sense of sympathy beyond kin and friends to collaborative partners. To coordinate their collaborative activities cognitively, early humans evolved skills and motivations of *joint intentionality,* enabling them to form together with a partner a joint goal and to know things together with a partner in their personal common ground (Tomasello, 2014). On the individual level, each partner had her own role to play in a particular collaborative activity (e.g., hunting antelopes), and over time there developed a common-ground understanding of the ideal way that each role had to be played for joint success. These common-ground role ideals may be thought of as the original socially shared normative standards. These ideal standards were impartial in the sense that that they specified what either partner, whichever of us that might be, must do in the role. Recognizing the impartiality of role standards meant recognizing that self and other were of equivalent status and importance in the collaborative enterprise.

In the context of partner choice, in which all individuals had bargaining leverage, this recognition of self–other equivalence led to a mutual respect among partners. And since it was vital for partners to exclude free riders, there also arose the sense that only collaborative partners (and not free riders) were deserving of the spoils. The combined result was that partners came to consider one another with mutual respect, as equally deserving second-personal agents (see Darwall, 2006). This meant that they had the standing to form with one another a joint commitment to collaborate (see Gilbert, 2003). The content of a joint commitment was that each partner would live up to his role ideal and, further, that both partners had the legitimate authority to call the other to task for less than ideal performance. Early humans' sense of mutual respect and fairness with partners thus derived mainly from a new kind of cooperative rationality in which it made sense to recognize one's dependence on a collaborative partner, to the point of relinquishing at least some control of one's actions to the self-regulating "we" created by a joint commitment. This "we" was a moral force because both partners considered it legitimate, based on the fact that they had created it themselves specifically for purposes of self-regulation, and the fact that both saw their partner as genuinely deserving of their cooperation. Collaborative partners thus felt responsible to one another to strive for joint success, and to shirk this responsibility was, in effect, to renounce one's cooperative identity.

In this way, participation in joint intentional activities—engendering both the recognition of partners as equally deserving second-personal agents and the cooperative rationality of subordinating "me" to "we" in a joint commitment—created an evolutionarily novel form of moral psychology. This novel form of moral psychology was not based on the strategic avoidance of punishment or reputational attacks from "they" but, rather, on a genuine attempt to behave virtuously in accordance with our "we." And so was born a normatively constituted social order in which cooperatively rational agents focused not just on how individuals do act, or on how I want them to act, but, rather, on how they *ought to* act if they are to be one of "us." In the end, the result of all of these new ways of relating to a partner in joint intentional activities added up for early humans to a kind of *natural, second-personal morality.*

The second evolutionary step in this hypothesized natural history—beginning with the emergence of *Homo sapiens sapiens* some 150,000 years ago—was prompted by changing demographics. As modern human groups started becoming larger, they split into smaller bands that were still unified at the tribal level. A tribal-level group—call it a culture—competed with other such groups for resources, and so it operated as one big interdependent "we," such that all group members identified with their group and performed their division-of-labor roles aimed at group survival and welfare. Members of a cultural group thus felt special senses of sympathy and loyalty to their cultural compatriots, and they considered outsiders to be free riders or competitors and so not deserving of cultural benefits. To coordinate their group activities cognitively, and to provide a measure of social control motivationally, modern humans evolved new cognitive skills and motivations of *collective intentionality*—enabling the creation of cultural conventions, norms, and institutions (see Searle, 1995)—based on *cultural* common ground. Conventional cultural practices had role ideals that were fully "objective" in the sense that everyone knew in cultural common ground how anyone who would be one of "us" had to play those roles for collective success. They represented the right and wrong ways to do things.

Unlike early humans, modern humans did not get to create their largest and most important social commitments; they were born into them. Most important, individuals had to self-regulate their actions via the group's social norms, the breach of which evoked censure not only from affected persons but also from third parties. Deviance in a purely conventional practice signaled a weakness of one's sense of cultural identity, but violation of a moral norm—grounded in second-personal morality—signaled a moral breach (see Nichols, 2004). Moral norms were considered legitimate because the individual, first, identified with the culture and so assumed a kind of coauthorship of them and, second, felt that her equally deserving cultural compatriots deserved her cooperation. Members of cultural groups thus felt an obligation to both follow and enforce social norms as part of their moral identity: to remain who one was in the eyes of the moral community, and so in one's own eyes as well, one was obliged to identify with the right and wrong ways of doing things (see Korsgaard, 1996a). One could deviate from these norms and still maintain one's moral identity only by justifying the deviation to others, and so to oneself, in terms of the shared values of the moral community (see Scanlon, 1998).

In this way, participation in cultural life—engendering both the recognition that all in-group compatriots were equally deserving and a sense that the culture's collective commitments were created by "us" for "us"—created a second novel form of moral psychology. It was a kind of scaled-up version of early humans' second-personal morality in that the normative standards were fully "objective," the collective commitments were by and for all in the group, and the sense of obligation was group-mindedly rational in that it flowed from one's moral identity and the felt need to justify one's moral decisions to the moral community, including oneself. In the end, the result of all of these new ways of relating to one another in collectively structured cultural contexts added up for modern humans to a kind of *cultural and group-minded, "objective" morality.*

One outcome of this two-step evolutionary process beyond great apes—first to collaboration and then to culture—is that contemporary human beings are under the sway of at least three distinct moralities. The first is simply the cooperative proclivities of great apes in general, organized around a special sympathy for kin and friends: the first person I save from a burning shelter is my child or spouse, no deliberation needed. The second is a joint morality of collaboration in which I have specific responsibilities to specific individuals in specific circumstances: the next person I save is the firefighting partner with whom I am currently collaborating (and with whom I have a joint commitment) to extinguish the fire. The third is a more impersonal collective morality of cultural norms and institutions in which all members of the cultural group are equally valuable: I save from the calamity all other groupmates equally and impartially (or perhaps all other persons, if my moral community is humanity in general), with perhaps special attention to the most vulnerable among us (e.g., children). The coexistence of these different moralities—moral orientations or stances, if you will—is of course anything but peaceful. Conflicts among them are the source of many of the most perplexing moral dilemmas that humans face—should I steal the drug to save my friend? should I keep my promise if it means harm to unknown others?—that seemingly have no fully satisfactory solutions (Nagel, 1986). The bare fact of such unsolvable incompatibilities in the dictates of morality suggests a complex and not totally uniform natural history in which different cooperative challenges have been met in different ways at different times.

The possibility that humans operate with several different, sometimes incompatible, moralities—and that they are due, at least in part, to processes of natural selection—raises the specter, feared by many thoughtful persons from Darwin's time on, that evolutionary explanations may serve to undermine the whole idea of morality. But this need not be the case. The point is that the ultimate causation involved in evolutionary processes is independent of the actual decision making of individuals seeking to realize their personal goals and values. The textbook case is sex, whose evolutionary raison d'être is procreation but whose proximate motivation is most often other things. The fact that the early humans who were concerned for the welfare of others and who treated others fairly had the most offspring undermines nothing in my own personal moral decision making and identity. I am able to speak the English language only because of my evolutionary, cultural, and personal histories, but that does not determine precisely what I decide to say at any given moment. In all, we should simply marvel at the fact that behaving morally is somehow right for the human species, contributing to humans' unparalleled evolutionary success, as well as to each individual's own sense of personal moral identity.

And so with this apologia, let us tell a story, a natural history, of how human morality came to be, beginning with our great ape ancestors and their sympathy for kin and friends, proceeding through some early humans who began collaborating interdependently with one another with joint commitments and a sense of partner equivalence, and ending with modern humans and their culturally constituted social norms and an objectified sense of right and wrong.

2

Evolution of Cooperation

And could this scarcity [of resources] not be alleviated by joint activities, then the domain of justice would extend only to the avoidance of mutually destructive conflicts, and not to the cooperative provision of mutual benefits.

—DAVID GAUTHIER, *MORALS BY AGREEMENT*

Sociality is not inevitable. Many organisms live, for all practical purposes, completely solitary lives. But many other organisms live socially, prototypically as they stay in close proximity to others of their kind to form social groups. The evolutionary function of this grouping is primarily protection against predation. Such "safety in numbers" sociality is sometimes called cooperation, as individuals aggregate with others relatively peacefully. But in more complex social species, cooperation may manifest in more active social interactions, such as altruistic helping and mutualistic collaboration.

The increased proximity of social life brings with it increased competition for resources. In social species individuals must actively compete with one another on a daily basis for food and mates. This competition can even lead to physical aggression, which is potentially damaging to all involved, and thus to a system of dominance status in which individuals with lesser fighting ability simply allow those with greater fighting ability to have what they want.

We thus have the two fundamental axes of animal sociality (Figure 2.1): a horizontal axis of cooperation based in individuals' propensities (high or low) for affiliating with (or even collaborating with or helping) others of their kind, and a vertical axis of competition based in individuals' power and dominance (high or low) in contesting resources. Finding a satisfactory balance between cooperation and competition is the basic challenge of a complex social life.

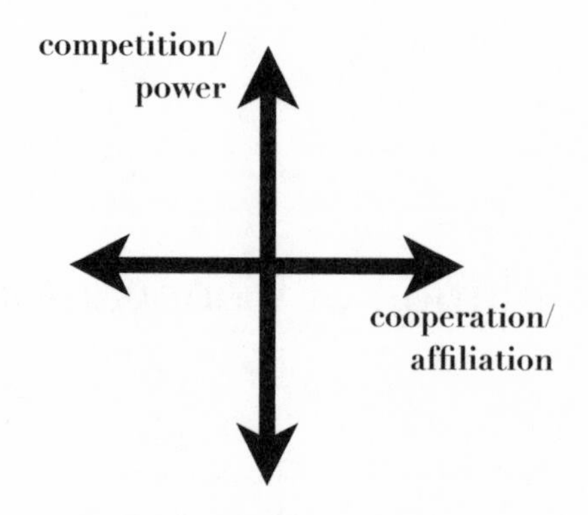

FIGURE 2.1. The two dimensions of social life for complex organisms.

In a Darwinian framework competition requires no special explanation, of course, but cooperation does. Acting in ways that benefit others is a stable evolutionary strategy only under certain conditions. The first task in this chapter, therefore, is to examine how cooperation works in evolution in general, using as an organizing theme the principle of interdependence. We then use this theoretical framework to characterize the nature of cooperation in great ape societies in particular, with the goal of characterizing, as a starting point for our natural history of human morality, the cooperative interactions of the last common ancestor of humans and other great apes some 6 million years ago.

Foundations of Cooperation

Cooperation presents a variety of puzzles for the theory of evolution by means of natural selection. We need not solve them all here. For current purposes all we must do is to identify those evolutionarily stable patterns of cooperation that are relevant for our investigation of the human species. In identifying these patterns we will be especially concerned both with the proximate (psychological) mechanisms—the cognitive, social-motivational, and self-regulatory processes—that enable the individuals of complex social species to cooperate with one another, and with the adaptive conditions under which those psychological processes might have come to be favored by natural selection.

Evolutionarily Stable Patterns of Cooperation

Standard evolutionary theory dictates that cooperation can be maintained as an evolutionarily stable strategy only if it is not overly detrimental to the reproductive fitness of the individuals involved (altruism is often defined by evolutionary biologists—humorously but pointedly—as "that which cannot evolve"). But there are a number of classic interactive categories that describe ways in which individuals may temporarily suppress their own immediate self-interest to cooperate with others, without thereby sacrificing themselves and their progeny out of existence in the long run. Following the theory of multi-level selection, it is most useful to explicate three broad categories, distinguished by the level at which they operate: kin selection operates at the level of the gene; group selection operates at the level of the social group; and mutualism and reciprocity operate at the level of the individual organism. Each of these categories of cooperative behavior may potentially be realized via a wide variety of different proximate mechanisms in different species.

First, perhaps the most basic process in the evolution of cooperation is kin selection. Darwin wondered why social insects, such as ants and bees, so readily sacrifice for one another (to the point that there even exist sterile helpers). In the context of modern genetics, Haldane and Hamilton solved the problem by noting that in social insects individuals living in the same social group share more genes with one another than do groupmates in other animal species. By helping others, individual ants and bees are promoting copies of their own genes; they are, in a sense, helping themselves. Dawkins (1976) pushed this view to the extreme, considering all of evolution from this "gene's eye view."

The proximate mechanisms for kin selection are normally quite simple. One has to be predisposed to do things that help others (without necessarily understanding cognitively that one is doing so), and one must direct this behavior selectively to kin. This selectivity toward kin is most often accomplished via spatial proximity. For example, ants and bees just do things that help others in their immediate environs, and even more cognitively complex organisms, such as humans, most often identify as kin those with whom they have grown up in close physical proximity (Westermarck, 1891). This psychological simplicity means that kin selection was not a likely breeding ground for the many complex cognitive distinctions and judgments underlying human morality. However, it almost certainly was responsible for the basic prosocial emotion of sympathy, originating in the context of parent–offspring bonding and helping one's kin. As we shall see in the account of great ape cooperation, some species then had occasion to extend their sympathy beyond kin to "friends."

A second major process in the evolution of cooperation is, controversially, group selection. The theory of group selection takes not a gene's eye view on the process but, rather, a group's eye view, with some theorists even noting that a multicellular organism is simply a group of cooperating single-cell organisms (Wilson and Wilson, 2008). The basic idea is that if the social groups of a species are internally homogeneous genetically, and these social groups are at the same time well differentiated from one another genetically, then these groups may actually become units of natural selection themselves. Cooperation enters the story because one can imagine that social groups with more cooperators outcompete social groups with more noncooperators. Individual cooperators are thus at a disadvantage within their group relative to noncooperators (who enjoy the benefits but do not pay the costs), but their group flourishes and so they have an advantage over individuals from other groups of the same species. Most theorists agree that group selection is possible in principle but that, in fact, in most cases there is too much gene flow between groups (due to immigration) for group selection to be a powerful force in more than just a few isolated cases.

Once again, the proximate mechanisms for group selection are simple. Again, one must simply be predisposed to do things that help others (without necessarily understanding cognitively that one is doing so), and one must direct this behavior selectively to groupmates, again recognized in most cases by spatial proximity.[1] Whereas group selection of this type may not have played a crucial role in the evolution of human cooperation and morality, a variant called cultural group selection almost certainly did, albeit quite late in the process. Cultural group selection does not primarily concern genetic evolution but, rather, cultural evolution, as individuals in a group conform to one another's behavior via social learning (thus promoting homogeneity of behavior), and even immigrants conform as well (thus solving the immigration problem). A secondary phase of gene–culture coevolution may also ensue such that individuals best able to, for example, socially learn have an adaptive advantage. Cultural group selection will play a key role in the later stages of our account of the evolution of human morality, as groups that are able to foster and encourage cooperation among their constituent members—through such things as social norms and institutions—outcompete neighboring groups that are not as good at these things.

Third, and at the crux of the current account because of their potential influence on psychological mechanisms, are processes of mutualism and reciprocity. These processes both work at the level of the individual organism, and both operate evolutionarily by somehow "paying back" the individual for his cooperation, either in the moment or later.

From an evolutionary point of view, processes of mutualism are easily explained because all cooperating individuals immediately benefit (although there can still be problems of free riding in some cases). For this reason, very little attention is paid to mutualism in the theoretical literature on the evolution of cooperation, human or otherwise (much more is paid to altruism). But, in fact, mutualistic collaboration is responsible for many of the most distinctive characteristics of human cooperative cognition and sociality. It is responsible in the sense that, early in human evolution, the need for specific types of mutualistic collaboration created the adaptive conditions within which an especially intricate and complex set of proximate mechanisms evolved for regulating social coordination and communication: shared intentionality (Tomasello, 2009, 2014). These proximate (psychological) mechanisms have been accorded little research attention in this context, but they are absolutely crucial—or so we will argue—for understanding the evolution of human cooperation and morality.

As for reciprocity, the classic version is so-called reciprocal altruism (Trivers, 1971): I help or defer to you on one occasion, and you reciprocate by helping or deferring to me on the next occasion, so that we both benefit in the long run. But how does this work psychologically? Classic tit-for-tat reciprocity (sometimes characterized as "you scratch my back, and I'll scratch yours") is often thought of implicitly as a kind of social contract in which we agree ahead of time to obligate ourselves to a future course of action. Although no one would seriously propose a social contract for nonhuman animals, without a social contract it is difficult to understand how reciprocity might work. The first problem is that reciprocal altruism has no explanation for the initial act of altruism at all, which has to be, in this account, blind optimism or accident. The second problem is the powerful incentive to defect: once you have given me my benefit, I have no incentive to give you yours in return; I should simply quit while I am ahead. My only incentive is that perhaps it will cause you to give me an additional benefit in return. But why should it? You have the same incentive to defect that I have. Without some kind of agreement, reciprocity has no real rational or emotional power to motivate altruistic behavior on its own. Indirect reciprocity brings reputation into the picture, but in the end it is also plagued by the same two problems of motivating the first act and cheating.[2]

There is no doubt that behavioral patterns of reciprocity occur widely in nature. The issue is the proximate mechanisms underlying them. What is needed, for current purposes at least, is a more psychologically realistic account to replace the implicit contract view of reciprocity. A good start is the typology of de Waal (2000). Most important, he distinguishes what he calls calculated reciprocity from emotional (or attitudinal) reciprocity. Calculated reciprocity is the implicit contract: we each keep track of who has done what for whom and stop cooperating if we are giving more than we are getting. Not surprisingly, this type of reciprocity seems to be very rare in nature. More frequent, especially in mammals with their propensity for forming long-term emotion-based social relationships, is emotional reciprocity. With emotional reciprocity, individuals form emotional bonds with those who help them (perhaps based on the mechanism by which offspring bond to those who succor and protect them), and then they naturally help those with whom they are socially bonded—kin and "friends," as it were. Emotional reciprocity would seem to be widespread in primates and other mammals, at the very least, but it raises the questions of why individuals form friendly social relationships with nonkin in the first place, why they help those friends, and how their friendships affect their reproductive fitness.

Interdependence and Altruism

Virtually all formal theories of the evolution of cooperation (e.g., Nowak and Highfield, 2011) conceptualize the individual as an asocial monad in constant competition with all other members of its species in a struggle to pass along its genes. But this view, while in some sense valid, is seriously incomplete in the case of cognitively and socially complex organisms, not to mention that it has little concern for proximate mechanisms. The main point is that cognitively and socially complex organisms are enmeshed in many and varied social relationships and interdependencies with others, and this means—assuming that these relationships and interdependencies are important to their fitness—that helping or cooperating with those others, reciprocally or otherwise, is not a sacrifice but an investment.

Consider Maynard Smith's (1982) famous hawk–dove interaction. Two individuals of a species, who know nothing about one another, approach a small patch of food. They could both cooperate (play "dove"), and each get half the food; but each is also tempted to chase the other away (play "hawk") and take it all—although this may end up in a destructive fight. They are in a prisoner's dilemma: to maximize food intake, the best strategy of each is to play hawk no matter what the other does, and the inevitable result is a destructive fight.[3] But now consider how things change if the individuals have an important social relationship. A male approaches the food at the same time as the only female in the group who will mate with him. Because his future reproductive success is entirely in her hands—he is 100 percent dependent on her for passing on his genes—he does not want her to go hungry. His preferred outcome is that they each get some food. If the female also depends on the male as a mating partner, then she does not want him to go hungry either. Now they are *inter*dependent, and there is no prisoner's dilemma because neither prefers the situation where he or she gets all the food and the other gets none. They care about one another's welfare.

The classic reciprocity account—as it is typically characterized in formal models—does not recognize such interdependencies; that is, it does not recognize the importance of social relationships for cooperative interactions. And, of course, different types of partners are differentially important. Accordingly, Roberts (2005) proposes what he calls the stakeholder model, in which individuals have a stake in the well-being of specific other individuals, for example, their mating or coalition partners. For an individual altruistic act,

$$sB > C,$$

where, as in Hamilton's famous equation for kin selection, B represents the reproductive benefits to the actor, which must exceed the costs to the actor, C, when the benefits are conditioned by the "stake," s, that the actor has in the recipient (analogous to Hamilton's coefficient of relatedness). The variable s simply represents how important it is to the actor that the recipient be alive and in good shape for future interactions. How specifically an individual identifies those on whom she is dependent and how dependent she is on them is a function of the particular cognitive mechanisms available to the particular species, varying from simple innate heuristics to complex learned judgments.

Note that the stakeholder model applies asymmetrically: it tells me whether I should help someone based on my stake in him. Whether or not he has a stake in me is irrelevant. Thus, I might have a stake in an alarm caller, since I rely on his performance to avoid predators. I should therefore help him as needed to keep him in good shape for his job. But of course he may be doing the job that nobody else wants to do precisely because I and others reward him for doing it. Now we have again a case of *inter*dependence, though in this case the two dependencies involved are of a different nature, and the relevant acts occur at different times. In more mutualistic activities, such as mating, cooperative hunting, or coalitionary quests for dominance, the same basic processes of interdependence are at work, only simultaneously and in a more symmetrical manner: we both benefit at the same time and in similar ways from our interdependent collaboration. Later, we will argue that an interdependent partnership in a mutualistic collaborative activity is a particularly important situation for both cooperation and altruism because of the symmetrical stability attained: each depends on the other in an immediate and urgent way that discourages cheating.

Individuals in socially complex species are dependent on, and interdependent with, many other groupmates in many different ways. Indeed, Clutton-Brock (2002) has argued that the essence of social life is interdependence (though he does not use this term) and so proposes a mechanism called group augmentation that applies to social life in general. If my prosperity depends on my social group (e.g., for defense against predators or other groups), then it is in my interest to keep each of my groupmates alive and prosperous. Social beings thus have at least a small stake in each of their groupmates. The outcome is that the overall stake I have in any given groupmate is the sum of many particular stakes I have in her, for example, as alarm caller, as coalition partner, as group member, and so forth. Thus, I will perform an altruistic act for her when

$$s_1B_1 + s_2B_2 \ldots s_kB_k > C,$$

where each different term represents one way in which I am dependent on her, have a stake in her, each at its own quantitative level. Again, organisms obviously do not need to cognitively compute their stake in others before acting; as always, Mother Nature provides cognitive heuristics and other shortcuts appropriate to the species, with more socially and cognitively sophisticated organisms presumably being more skillful and flexible in their determinations.

In the stakeholder model, then, one could say that individuals do get "paid back" for their niceness, so one could call it something like "pseudo-reciprocity" (Bshary and Bergmueller, 2008). Fine. But the important point is that *unlike classical reciprocity, the altruist's behavior is not contingent on the recipient responding or being influenced by the help in any way (or on the altruist's anticipation of any such responding or influence).* The recipient will simply continue doing what he always does—alarm calling, mating, being a coalition or hunting partner, or being in the social group—because it is in his interest to do so, and this just happens to benefit the altruist, as a by-product, as it were. A more active way of thinking about the altruist's behavior, then, is as a kind of investment in the recipient; she invests in his well-being since that contributes to her well-being (Kummer, 1979). In this view, emotional reciprocity is most accurately characterized as mutual investments among interdependent friends, who help one another not in order to pay back past acts but in order to invest in the future. In some cases, each individual may be dependent on the other precisely because of the benefits he provides—for example, they share food with one another reciprocally—but from a proximate point of view the altruistic act is not motivated by any specific previous act, only by the goal of maintaining the relationship. We might thus conceptualize the situation as individuals living *symbiotically* (a concept typically applied only to interactions between species), as there is no exchange of favors or anything of the kind, only individuals going about attempting to increase their fitness directly.

This way of looking at things does not have the many problems of reciprocity—specifically, the problem of motivating the first altruistic act or the problem of defecting—because there is no direct contingency of altruistic acts (although, of course, over time a relationship may break down for many reasons). One individual helps another do what she would do in any case, for her own reasons—up to a mathematical point. But this account does, of course, still have a potential problem of free riding because one can lag on the helping: it would be best if someone else helped my alarm-calling groupmate so that I could get the benefits without paying any costs. But as Zahavi (2003) has pointed out, the exact same logic applies to kin selection: it is in my interest to help my sibling because he shares my genes, but my first preference is that someone else help him so that I do not have to bear the costs or risks. And so, of course, interdependence as described by the stakeholder model does not solve all of the problems of cooperation in one fell swoop; rather, using the same logic as kin selection, it changes the cost-benefit analysis significantly.

The interdependence perspective thus integrates mutualism and reciprocity in a natural way, and it motivates reciprocity in a much more stable way than classical accounts. It also places altruism in a new light. Altruism is not an improbable achievement against the individualizing forces of natural selection; rather, it is an integral part of the social lives of all beings that live with others interdependently—up to a (mathematical) point. Everyone helps and gets helped, up to a point, because everyone is important to someone in some way, up to a point. This view also accords well with the prescient views of Kropotkin (1902) on "mutual aid" as playing a crucial role of the everyday lives of social beings who must struggle more against the exigencies of the physical environment (sometimes cooperatively) than against one another.

Partner Control, Partner Choice, and Social Selection

Cooperators do best when they are surrounded by other cooperators. So once the individuals of a species have begun down the cooperative path, they may actively attempt to influence others around them in a cooperative direction. They may do this most directly through acts of so-called partner control—most often the punishment of noncooperators—which may be seen as the kind of opposite of investing positively in cooperators and friends through acts of helping. The problem with punishment is that it is costly, or at least risky, for the punisher, for example, if the individual being punished rebels. A safer alternative, if it is available, is so-called partner choice in which cooperators simply avoid interacting with cheaters. Although there may be simple ways for avoiding cheaters in some cases, in more socially complex organisms partner choice often requires sophisticated judgments, based on past experience, about which of several individuals will make the best partner. In all, partner control and partner choice represent important complements to the process of investing positively in those on whom one depends. In these cases, individuals actively attempt to influence those on whom they depend, or might depend, either by coercing bad partners into being good ones or by choosing their partners wisely.

Given processes of either partner control or partner choice, over time what emerges is what West-Eberhardt (1979) calls social selection. In Darwin's (1871) modified account of organic evolution, the process of natural selection is complemented by the process of sexual selection. Sexual selection is not a totally new evolutionary process; it is just that in this case the selection is being performed not by the physical environment (as in classical natural selection) but, rather, by the social environment. In sexual selection, individuals of the opposite sex choose potential mates based on characteristics indicating such things as health, strength, and fecundity (e.g., large size, bright coloration, or youth). Those characteristics are thus selectively favored for purposes of mating, and this increases the reproductive fitness of those possessing them.

Social selection is simply a generalization of this process. Individuals in the social group may favor other individuals for all kinds of reasons in addition to sexual attractiveness, and this may affect both the recipient's survival and his reproductive success. Thus, if the individuals of a social group do the most beneficial things for the best alarm callers, then the characteristics of good alarm callers—keen perception, fast reactions, loud calling—will be socially selected for in alarm callers. If individuals of a social group need grooming partners, then individuals who are enthusiastic and good groomers will be selected for, along with their special characteristics. In some interactions, one individual may have more "leverage" than others in the sense that he is a more important partner; for example, dominants might be expected to be more in demand as coalition partners than subordinates, so they can demand more of their potential partners than can subordinates. We may thus think of the kind of socially complex decision making that occurs in partner control and partner choice as a kind of "biological market" (Noe and Hammerstein, 1994).

Although in principle virtually any physical or behavioral characteristic may be subject to social selection, for current purposes cooperation is a special case. If we assume, for example, that in some species collaborating with others to obtain food brings mutualistic benefits to all involved, then one can imagine a biological market based on partner choice and control in which what is being socially selected is the characteristics of good cooperators, for example, tolerance for partners in feeding situations, skills at coordinating and communicating with partners, propensities for helping partners as needed, and tendencies for shunning or punishing free riders—and what is being selected against, of course, are the characteristics of cheaters and incompetents.

Summary

To prepare for our upcoming natural history of human morality, we have briefly considered all three levels of multilevel selection. Kin selection was undoubtedly critical in primate (even mammalian) cooperation, well before humans emerged, in building up the emotional substrate for protecting and caring for offspring, which was then in some cases co-opted for protecting and caring for friends. Cultural group selection—as a special instance of group selection without its attendant problems—very likely played an important role near the end of the process, as modern human cultural groups competed with one another for resources and the most cooperative (or moral) groups won out.

But in our hypothesized natural history the main action will be at the level of the individual organism. This is because, in the current account, acting morally means interacting with others cooperatively by means of and through certain psychological processes. What we have done so far is simply to reconceptualize the evolutionary processes of cooperation that occur at this individual level, most especially mutualism and reciprocity. We have proposed that what is most basic at this level are dependencies (symbioses) among individuals, which may produce mutualistic or reciprocal patterns of cooperation by any of many different proximate mechanisms (e.g., emotional reciprocity in mammals). They also motivate individuals to care for or invest in those on whom they depend (altruism) and to attempt to make their partners as cooperative as possible (partner choice and control). This reconceptualization of individual-level cooperative interactions and relationships lays important theoretical groundwork for explaining the evolutionary emergence of human morality.

Great Ape Cooperation

Given this theoretical framework in which cooperation is based mainly on the principle of interdependence, we may now begin our natural history of human morality proper. We do this by attempting to characterize, as best we can, the social lives of the last common ancestor of humans and other great apes, who lived somewhere in Africa approximately 6 million years ago. We use as contemporary models the social lives of humans' nearest living relatives, the great apes, especially those of humans' very nearest living relatives, chimpanzees and bonobos (although in actuality the vast majority of both field and experimental research is with chimpanzees). We look first at several aspects of their social interactions with conspecifics in the wild, and then at their behavior in experiments directly testing for senses of sympathy and fairness.

Sociality and Competition

Chimpanzees and bonobos live in highly complex social groups typically comprising several dozen or scores of individuals of both genders (so-called multimale, multifemale groups). Daily life is structured by a fission–fusion organization in which small parties of individuals forage together for some time, only to disband soon thereafter in favor of new parties. Males live their whole lives in the same group in the same territory; females emigrate to a neighboring group during early adolescence. During their development, individuals form with others various kinds of long-term social relationships. Most important is, of course, kinship, but also important are relationships with nonkin based on dominance and something like friendship. Much of the complexity of chimpanzee and bonobo social interaction results from the fact that they also recognize and react to these same social relationships as they occur among third parties in the group. Interactions between neighboring groups are almost totally hostile for chimpanzees, whereas for bonobos interactions with foreigners are more peaceful.

Both chimpanzees and bonobos compete with groupmates all day every day. This is not only in the indirect evolutionary sense of competing to pass along genes but also in the more immediate sense of competing face-to-face for food, mates, and other valuable resources. For example, in the case of food, the prototypical situation is that a handful of individuals travel until they find a fruiting tree. Each individual then scrambles up the tree on its own, procures some fruit on its own, and seeks maximum spacing from others to eat. Such scramble competition, in which the winner is the one who gets there first, is often complemented by contest competition, in which the winner is the one who wins the fight or dominance contest, for example, when a dominant in the tree takes what it wants while nearby subordinates defer. Both chimpanzee and bonobo males compete, and even fight, for access to females, though this competition is clearly more intense in chimpanzees. Interestingly, many agonistic encounters in both species are not about immediate access to a resource but, rather, about one individual simply asserting dominance over another—which will then translate into easier access to resources of various kinds in the future. Contests over dominance, therefore, act as a kind of proxy for contests over resources.[4]

Cognitively, chimpanzees and bonobos are built for competition. Thus, not only are they intentional, decision-making agents, who make instrumentally rational decisions themselves, but they also perceive others as intentional, decision-making agents with whom they must compete. They understand that the behavioral decisions of their competitors are driven both by their goals (what they want to be the case) and by their perception of the situation (what they perceive to be the case)—a kind of perception–goal psychology (Call and Tomasello, 2008). As just one example, a subordinate individual will go for a piece of food only if it predicts that the nearby dominant will not go for it, either because (1) he does not want it or (2) he does not see (or has not seen) it (Hare et al., 2000, 2001). A number of studies have documented that great apes use social-cognitive skills such as these more readily in competitive than in cooperative or communicative contexts (Hare and Tomasello, 2004; Hare, 2001), a kind of "Machiavellian intelligence" (Whiten and Byrne, 1988). Great apes are especially skillful compared with other primates at manipulating the physical world to their benefit through the use of tools, and the social world to their benefit through the use of flexible communicative gestures (Call and Tomasello, 2007). In all, great ape social cognition seems built for outcompeting others by outsmarting them.

Obviously, individuals who live in social worlds structured by dominance must learn to control their impulses for immediate self-gratification. In their daily lives, chimpanzees and bonobos exercise self-control constantly, as subordinates inhibit their impulses to do things like take food or pursue mates if it will get them into trouble with a dominant. In systematic experimental tests, chimpanzees have shown that they can (1) delay taking a smaller reward so as to get a larger reward later, (2) inhibit a previously successful response in favor of a new one demanded by a changed situation, (3) make themselves do something unpleasant for a highly desirable reward at the end, (4) persist through failures, and (5) concentrate through distractions. They do all of these things at roughly the level of three-year-old human children, and at a lower level than six-year-old human children (Herrmann et al., in press). Chimpanzees' skills of impulse control, self-control, emotion regulation, and executive function—as these skills are variously called—are thus clearly sufficient for inhibiting selfish impulses in deference to others when it is prudent to do so.

It is also noteworthy in this connection, and perhaps of special relevance for the evolution of morality, that chimpanzees and bonobos would both seem to have and to recognize in others basic emotions such as fear, anger, surprise, and disgust. Although there is not much systematic research on great ape emotion recognition, in their natural social environments individuals clearly avoid others displaying anger, look around fearfully when others display fear, and investigate things when others display surprise. Experimentally, Buttelmann et al. (2009) found that when a human looked into one bucket and showed disgust, and into another bucket and showed happiness/pleasure, chimpanzees preferentially chose to have whatever was in the more pleasure-inducing bucket.

Collaboration for Competition

In all, then, chimpanzees' and bonobos' social lives are most immediately and urgently structured by a matrix of social competition for food, mates, and other resources. They cooperate in some ways as well, of course, but in almost all cases the key to understanding their cooperation is this same overarching matrix of social competition. Thus, in their review of the major dimensions of chimpanzee social life, Mueller and Mitani (2005, p. 278) observe: "Competition . . . frequently represents the driving force behind chimpanzee cooperation."

Chimpanzees, bonobos, and other great apes regularly collaborate with conspecifics in two key contexts. First, like many mammalian species, great apes engage in various forms of group defense as they compete with neighboring groups for resources and territory. In these agonistic encounters great apes essentially "mob" their enemies to drive them away, as do many animal species, and this does not require them to collaborate in any especially sophisticated ways. Notably in this context, small groups of male chimpanzees actively "patrol" their border, engaging agonistically with any individuals from neighboring groups that they encounter (Goodall, 1986). Presumably, acts of group defense are a reflection of groupmates' interdependence with one another, at the very least as a need for group augmentation, but more urgently in protecting those specific others on whom they depend in various ways.

Second, and again like many mammalian species, chimpanzee and bonobo individuals attempt to improve their odds for success in intragroup agonistic encounters by forming coalitions with others (Harcourt and de Waal, 1992). These aggressive encounters are most often over dominance status itself, presumably as proxy for priority in obtaining resources (e.g., when a coalition simply forces an individual to move from his current resting spot). Whereas in many monkey species it is typically kin who support one another in fights (e.g., rhesus monkey matrilines), in chimpanzees it is mostly nonkin (Langergraber et al., 2007). This cooperating in order to compete requires individuals to monitor simultaneously two ongoing social relationships, but again, the coalition partners do not do anything special to coordinate their actions other than fighting side by side. And again as in many mammalian species, great ape combatants often actively reconcile with one another after fights, presumably in an attempt to repair the social relationship on which they both depend (de Waal, 1989a). In some cases, dominant males may intervene to break up interactions in which coalitions (which may later challenge him) are beginning to form (de Waal, 1982).

Coalitionary support is crucial in chimpanzee and bonobo dominance contests; it pays to have good and powerful friends. Therefore, individual chimpanzees and bonobos cultivate friends, quite often through reciprocal coalitionary support (de Waal and Luttrell, 1988). They also cultivate friends through other affiliative behaviors, such as grooming and food sharing. Thus, much evidence suggests that grooming in chimpanzees is preferentially directed to potential coalition partners (see Mueller and Mitani, 2005, for a review), and over time individuals that have been preferentially groomed by a partner preferentially groom that partner as well (Gomes et al., 2009). In addition, Mitani and Watts (2001) have shown that male chimpanzees preferentially "share" (i.e., for the most part, tolerate the taking of) meat and other food with their coalition partners. Add to this the finding of de Waal (1989b) that individuals that groom one another also share food with one another preferentially, and the result is a relatively tight set of reciprocal relations among the golden triad of grooming, food sharing, and coalitionary support.[5] Importantly, there is very little evidence that great apes engage in any reciprocity of favors with individuals other than long-term social partners. In the only relevant experimental study, Melis et al. (2008) found that randomly paired chimpanzees did not preferentially help an individual that had just helped them over one that had not. They concluded that, despite clear evidence of long-term reciprocities, "models of immediate reciprocation and detailed accounting of recent exchanges (e.g., tit for tat) may not play a large role in guiding the social decisions of chimpanzees" (p. 951).

Interrelations within the golden triad are often interpreted as instances of reciprocity, and of course, on a purely descriptive level, they are. But de Waal (2000), as noted above, identifies several different proximate mechanisms that might underlie these observed behavioral patterns. And what de Waal and other investigators believe is underlying them is not "calculated reciprocity" involving mental scorekeeping but, rather, "emotional reciprocity," in which individuals simply develop positive affect toward those that help or share with them (see Schino and Aureli, 2009). In the current theoretical framework, what is happening is that individuals help or share with those for whom they feel the prosocial emotion of sympathy—that is, those on whom they are dependent—and the recipients then help or share with them because they now feel sympathy for them (based on their newly established dependence on them). This interdependence—sympathy in both directions—thus generates reciprocal patterns of helping and sharing in friendship interactions (see Hruschka, 2010, on friendship in evolutionary perspective). The key point for current purposes is that *great ape patterns of reciprocity on the behavioral level are underlain not by any kind of implicit agreement or contract for reciprocity, much less by any kind of judgments of fairness or equity, but only by interdependence-based sympathy operating in both directions.* Such sympathy, we will claim, is the one and only prosocial attitude of great apes and other primates, and it emerges only in relatively long-term social relationships.

Although there is not much systematic research on the topic, there would seem to be a good bit of partner choice in chimpanzee and bonobo coalition formation. In particular, as just noted, individuals seem to stick with, and even to cultivate as friends (through grooming and food sharing), good coalition partners. But there is an irony here: as individuals go about choosing their friends-cum-coalition partners, processes of social selection are, in an important sense, working against the evolution of cooperation. Since the coalition itself involves very little coordination—each partner simply does her individual best in the fight as they act in parallel—the best coalition partners are simply those that are dominant in fights, full stop. What partner choice in coalitions is socially selecting for most directly, then, are the characteristics of good and dominant fighters, not of cooperative and helpful partners.

The general picture is thus that the overarching context for cooperation and friendships in chimpanzees and bonobos is coalitionary competition for dominance, food, and mating opportunities. This is especially true in male chimpanzees, where such competition occurs most frequently, but it is also true of female chimpanzees and bonobos. For example, Williams et al. (2002) found that chimpanzee females preferentially forage with friends at least partly because friends are less likely than nonfriends to contest their acquisition of resources. Mueller and Mitani (2005, p. 317) summarize the situation as follows:

> The most prevalent forms of cooperation among chimpanzees . . . are rooted in male contest competition. Chimpanzee males maintain short-term coalitions and long-term alliances to improve their dominance status within communities and defend their territories cooperatively against foreign males. Other prominent cooperative activities, such as grooming and meat sharing, relate strategically to these goals. Females are far less social than males, and they do not cooperate as extensively. Nevertheless, the most conspicuous examples of female cooperation also involve contest competition, as females sometimes cooperate to kill the infants of rivals.

Overall, then, chimpanzees and bonobos live their lives embedded in constant competition for resources, so they are constantly attempting to outcompete others by outfighting them, outsmarting them, or outfriending them.

Collaboration for Food

The coalitions and alliances of great apes are aimed at procuring the same zero-sum resources for which individuals are competing. That is, individuals compete with one another for access to resources such as mates or food, and then coalitions of individuals join into this same competition for these same resources. Coalitions and individuals are thus fighting for pieces of one and the same foraging pie. But there is one prominent exception to this zero-sum competition. In some but not all chimpanzee groups, males hunt in small social parties for monkeys (although less frequently, bonobos apparently hunt in small parties for monkeys and other small mammals as well; Surbeck and Hohmann, 2008). Individuals cannot typically capture these monkeys on their own, and so the collaboration, in effect, expands the foraging pie.

In terms of coordination, in some cases (e.g., sites in eastern Africa with little forest canopy) the hunt resembles a kind of helter-skelter chase in which multiple individuals spy and then chase a monkey in parallel. In other cases, however (e.g., the Taï Forest in western Africa and Ngogo in eastern Africa), the forest canopy is continuous and the monkeys are quite agile, so such haphazard chasing will not succeed. Here the chimpanzees must, in effect, surround a monkey in order to capture it, requiring individuals to coordinate with one another (Boesch and Boesch, 1989; Mitani and Watts, 2001). Despite this, what is likely happening is a kind of individualistic coordination: each hunter is attempting to capture the monkey for itself (since the captor will get the most meat), and they take account of the actions of others in order to do so. Thus, one individual begins the chase, and then others go to the best remaining locations in anticipation of the monkey's attempted escape. The participants are not working together as a "we" in the sense of having a joint goal and individual roles within it; rather, they are operating in what Tuomela (2007) calls "group behavior in I-mode" (Tomasello et al., 2005; see Chapter 3). This means that, typically, the captor chimpanzee will steal away with the carcass if it can. But typically it cannot, and so all participants (and many bystanders) will get at least some of the meat by begging and harassing the captor (Boesch, 1994; Gilby, 2006).

Chimpanzee and bonobo individuals are clearly interdependent with one another during the hunt itself. Indeed, experiments have shown that chimpanzees in similar situations understand that they need a partner to be successful (Melis et al., 2006a; see also Engelmann et al., in press). But—and this is crucial for our larger story in which humans become obligate collaborative foragers—chimpanzee and bonobo individuals do not depend on the group hunting of monkeys to survive. Most of their food comes from fruit, other vegetation, and insects. In fact, and perhaps surprisingly, chimpanzees hunt most often for monkeys not in the dry season when fruit and vegetation are scarcer but, rather, in the rainy season when fruit and vegetation are more abundant (Watts and Mitani, 2002). This is presumably because spending time and energy in a monkey hunt for an uncertain return makes most sense when there are plenty of backup alternatives if it fails. The group hunting of chimpanzees and bonobos is thus interdependent in the moment, but the participants are not interdependent with one another for obtaining food in general.

The degree to which chimpanzees and bonobos in the wild may actively choose collaborative partners for hunting—potentially creating situations of partner choice and social selection—is unclear. Melis et al. (2006a) found that after a fairly small amount of experience with one another, captive chimpanzees knew which individuals were good partners for them (in the sense of leading to collaborative success and so to food), and they chose those partners in preference to others. It is not clear, however, whether partner choice of this type happens in the wild, since hunting in the wild is mostly opportunistic and by preconstituted traveling parties; there is little opportunity for choice. And if there are calls for recruiting others to the hunt, as suggested by Crockford and Boesch (2003), these do not represent partner choice because they are not selectively aimed at good cooperators but, rather, at anyone. There would also seem to be nothing like partner control at work in chimpanzee group hunting either, as free riders are not actively excluded from the spoils (Boesch, 1994).

Collaboration of this type in the acquisition of food is rare in primates and other mammals. Social carnivores such as lions and wolves engage in group hunting, but it works mainly because the carcass of the prey is too large for any individual to monopolize afterward, and so everyone eats their fill with no need for active sharing (and no special efforts to exclude free riders). Social carnivores are clearly evolved for such group hunting, as for them it is obligate (there are no backup options). It seems unlikely, however, that chimpanzees and bonobos are evolved for the group hunting of monkeys since not all groups do it, only males do it, and even males do it only when forced to do so by the structure of the forest canopy. And backup foraging alternatives are always available. In experiments, when given the option of obtaining food on their own or collaboratively with a partner, chimpanzees are indifferent and simply choose the option where there is the most food for them (Bullinger et al., 2011a; Rekers et al., 2011). Additional evidence that chimpanzees and bonobos are not specifically adapted for collaborative foraging—they are collaborating based on general cognitive abilities—is presented in Chapter 3, where they are systematically compared with human children on a variety of cognitive and motivational propensities for collaboration.

In all, then, although chimpanzees' and bonobos' group hunting of monkeys and other small mammals is spectacular testament to their cognitive flexibility and intelligence, in general their abilities to create new resources by collaborating are limited. In the words of Mueller and Mitani (2005, p. 319):

> In comparison with humans, . . . the general lack of cooperative behavior by chimpanzees in noncompetitive contexts, such as foraging, is conspicuous. Cooperative food gathering occurs routinely among all human foragers (e.g., Hill, 2002). Even the simplest forms of such behavior, such as Hadza men climbing baobab trees to shake down fruits for the women below . . . lack an apparent equivalent in chimpanzee behavior.

Sympathy and Helping

Because our primary focus is on morality, a key question in all of this is whether chimpanzees and bonobos are doing anything altruistic in any of these cooperative activities, in the sense that they have as their proximate goal some benefit for the other with nothing but a cost for themselves. (For economists, the question is phrased as whether they have any "other-regarding preferences," whereas for moral psychologists the question is phrased as whether they have any sympathetic concern for the other.) The observational/correlation studies that have established the golden triad of grooming, food sharing, and coalitionary support are not able to answer this question definitively, nor are the naturalistic observations of group hunting. For example, grooming is rewarding not only to the one being groomed but also to the groomer, who usually procures some fleas in the process. Food sharing as "tolerated theft" or "sharing under pressure" is aimed most directly at keeping the peace. When individuals participate in a coalitionary contest, each participant stands to gain in dominance status with a win. And in group hunting, each hunter is hoping to capture the monkey for itself, or at least to get some pieces of meat by begging and harassing the captor. The methodological point is simply that naturalistic observations and correlational analyses by themselves are not sufficient for determining whether some observed behavior is other-regarding or sympathetically motivated.

But recently a number of experimental studies have investigated chimpanzees' tendency to help others in instrumental contexts. Methodologically, the key is that these studies have control conditions ensuring that the ape subject does not have other motives for its acts. Typically in these studies the helping act is not especially costly, other than energetically, and occurs in situations in which competition for food or other resources is absent. For example, in the first such study, Warneken and Tomasello (2006) found that three human-raised chimpanzees fetched an out-of-reach object for a human caretaker who was trying to reach it, but they did not fetch objects in identical situations in which the human had no use for it. Because human-raised chimpanzees helping their human caretakers might be a special situation, Warneken et al. (2007, study 1) followed up with a study in which they presented captive chimpanzees living in a seminatural setting with similar tasks involving an unfamiliar human. The chimpanzees helped these humans too (more than in control conditions), even paying an energetic cost to do so by climbing a few meters high to fetch the desired object. In another variation, the apes helped a human with food in her hand no more often than a human with empty hands, undermining the interpretation that their helping act was aimed getting the human to reward them.

But still, helping humans is not the perfect test. Therefore, Warneken et al. (2007, study 2) also tested chimpanzees in situations in which they could help a conspecific. In particular, when one individual was trying to get through a door, another (from an adjoining room) could pull open a latch for her. Chimpanzees reliably helped a conspecific in this situation, even though in control conditions they did not open the latch if the first chimpanzee was not trying to get through the door (or if it was trying to get though a different door). In another experimental paradigm, Melis et al. (2011b) found that chimpanzees even helped a conspecific to get food. That is, individuals who had no possible access to a piece of food released a hook to send it down a ramp to a conspecific in another room across the hall, which they did not do if there was no one at the other end of the ramp. Yamamoto and Tanaka (2009) observed that chimpanzees would give tools to others that needed them to rake in food, but not give them tools if they did not need them. And Greenberg et al. (2010) found that chimpanzees helped others get food even when there was no active solicitation from the one needing help (if the need was otherwise obvious). Arguably, the key in all of these experiments is that it was clear to the helper that it could not obtain food—the helping was not costly in terms of resources.

In a different experimental paradigm, Silk et al. (2005) and Jensen et al. (2006) gave chimpanzees the option of pulling one rope that delivered food on a board only to themselves or pulling a different rope that delivered that same amount of food both to themselves and, in addition, at the other end of the board, to a conspecific. Chimpanzees pulled the two ropes indiscriminately. But most likely chimpanzees' failure to help the conspecific get food in this experimental paradigm was due to their concentration on getting food for themselves. Either they were so excited that they ignored the other chimpanzee, or the food on both sides of the board put them in a food-competition mode that inhibited helping. In either case, the negative findings in this situation do not negate the numerous positive findings of instrumental helping in other situations.[6]

Instrumental helping would seem to be aimed at benefiting another individual at a small but nevertheless real (energetic) cost. The evolutionary function of instrumental helping may very well be the same as grooming and food sharing: to cultivate friends, especially as coalition partners. We cannot tell from these studies, as the recipient of the helping has never been systematically manipulated to see the effect on the helper. Nevertheless, what these studies do show—especially clearly because of their control conditions—is that chimpanzees in some situations do act with the proximate goal of benefiting the other. Warneken and Tomasello (2009) provide further evidence for this interpretation by comparing these chimpanzee studies with studies of human children in similar situations, and in every case the two species behave in very similar ways (although, as we shall see in Chapter 3, human children have been studied in a variety of other interesting contexts that chimpanzees have not). Another potentially relevant observation is that chimpanzee individuals not involved in a fight may "console" the loser after a fight (although the function and mechanism of this consoling behavior are both unclear; Koski et al., 2007).

One might infer from chimpanzees' helping behavior an underlying emotion of sympathy for the plight of the other who cannot obtain the food or tool it desires. There is no evidence one way or the other for this motivational interpretation in these experimental studies, but there are other recent studies that provide highly suggestive evidence that helpers sympathize with those whom they are helping. Crockford et al. (2013) found that during a grooming bout the chimpanzee doing the grooming (as well as the one receiving the grooming) shows an increase in the mammalian bonding hormone oxytocin. And Wittig et al. (2014) found similarly that the chimpanzee that gives up food during a food-sharing episode (which, admittedly, happened only rarely in their observations) also shows an increase in oxytocin. When one puts together these hormonal studies from the wild and the experiments on instrumental helping, this would seem to constitute very good evidence that chimpanzees feel sympathy for those whom they are helping.

In all, there is no reason to believe that these acts of helping are anything other than the genuine article. When costs are small, and food competition is absent, great apes help others. And there is at least some evidence—including acts of helping from other mammalian species such as rats (Bartal et al., 2011)—that underlying these altruistic acts is the oxytocin-based social emotion of sympathy. Of course, the evolutionary bases for this propensity to help others lie in some kind of payback; that is the logic of natural selection. But our focus here is on the proximate mechanisms involved, and in this case our best evidence is that chimpanzees and other great apes have a proximate motivation of sympathy for others in need, and this leads them to help, if the costs are not too great.

No Sense of Fairness

The other central dimension of morality, in addition to sympathy, is the sense of fairness or justice, and so we may also ask whether great ape cooperation exhibits anything substantive on this dimension. There are no experimental studies with great apes of so-called retributive justice in which one individual seeks to balance the books behaviorally with another via "an eye for an eye" retribution (although de Waal [e.g., 1982] reports some anecdotes of what he calls "revenge"). The closest is the study of Jensen et al. (2007), in which individuals that had food stolen from them prevented the thief from consuming it. But there are many interpretations of this behavior not having to do with fairness or justice. In contrast to this paucity of evidence for retributive justice, in the case of distributive justice—how resources are to be distributed among individuals—there are two sets of experimental results.

The first set of findings come from the famous ultimatum game. In this economic game, studied extensively in human adults, a proposer proposes a split of resources with a responder. The responder can accept the proposal, in which case they each keep their allocation, or reject it and then no one gets anything. In many experiments, human responders routinely reject low offers—say, 2 out of a total of 10—even though it means that they get nothing. The most likely reason for this "irrational" behavior is the responder's judgment that the offer is unfair. He is not going to go along with it—he is not going to be taken advantage of in this way—even if his rejection costs him resources. Quite often responders are even angry with unfair proposers. An adapted version of the ultimatum game has been given in nonverbal form to chimpanzees in two studies (Jensen et al., 2007; Proctor et al., 2013) and to bonobos in one study (Kaiser et al., 2012). In all three studies the result was identical: subjects virtually never rejected any nonzero offers. Presumably they did not because they were not focused on anything like the fairness of the offer, only on whether or not it would bring them food.

The other set of relevant findings come from studies of social comparison. The phenomenon is that humans are happy to receive X number of resources, unless they see others getting more, in which case they are unhappy. Presumably, again, the explanation for their unhappiness is a sense of being treated unfairly. Brosnan et al. (2005, 2010) report two studies claiming that chimpanzees will reject food given to them by humans (which they otherwise would accept) if they see another chimpanzee getting better food for the same or less effort. But when appropriate control conditions are run, this effect goes away. The necessary control conditions are those in which the chimpanzee just expects better food in the same context, which makes the offered food less appealing. Thus, Bräuer et al. (2006, 2009) replicated the Brosnan et al. studies with control conditions of this type (and also counterbalanced for order of conditions, which Brosnan et al. did not do) and found that chimpanzees did not reject food based on a comparison with what another chimpanzee obtained. They rejected food when they were frustrated at not getting a better food. This experimental paradigm thus does not show social comparison in chimpanzees, only food comparison. These negative results were corroborated, in a different paradigm, by Hopper et al. (2013): "Overall, the chimpanzees' responses appeared to be primarily influenced by the quality of the rewards they received, and not in relation to their partner's rewards." The same negative results have been reported for orangutans by Brosnan et al. (2011) and also by Bräuer et al. (2006, 2009), who in addition report inconclusive results with a small sample of bonobos.[7]

There is thus no solid evidence that great apes have a sense of fairness in dividing resources, and much evidence that they do not. (Although we should keep in mind that the ultimatum game is a demanding experimental paradigm for nonlinguistic creatures, and the social comparison studies are demanding in requiring subjects to reject otherwise palatable food.) Another potential source of evidence is great apes' emotional reactions to events that could potentially be perceived as unfair. As is discussed more fully in Chapter 3, the prototypical "reactive attitude" to unfairness is resentment. One reason that the findings of Brosnan et al. (2005, 2010) have seemed so compelling lies in the videos the authors often show of a monkey angrily throwing rejected food out of its cage. Given our analysis of these studies, it is highly unlikely that this emotional expression constitutes resentment that a conspecific is, unfairly, receiving a better piece of food. But it may be expressing anger at the human experimenter for handing over a poor piece of food when she could have handed over a good one. If this is true, then the emotional expression would constitute anger with a special social content, perhaps something like "I am angry that you are treating me without sympathy." Roughley (2015) indeed believes that anger with this kind of content may reasonably be called resentment*, where the asterisk indicates that it is not resentment proper but nevertheless a step on the way. It is a step on the way both because the content includes the prosocial emotion of sympathy and because it assumes some kind of social relationship, for example, friendship. I am angry not just at things in general, or at some animate thing over there, but at my friend with whom I have a relationship (even if it is interspecific) for treating me in this bad way. To avoid confusion, let us call this social anger.

The upshot is that, whereas the prototypical reactive attitude of resentment is expressed by one person to another person to protest unfair (or otherwise underserved) treatment, it is possible that apes experience something more like social anger. This would represent an initial step transforming the general mammalian emotion of anger into a specifically social form of anger for being treated badly (i.e., without sympathy) by someone with whom one has a valuable social relationship. True resentment would await the emergence of new types of moral judgments concerning deservingness, fairness, and respect.

Kin- and Friend-Based Prosociality

To summarize, as in virtually all social species, great ape individuals are dependent on their groupmates for surviving and thriving, but at the same time they compete with them for resources. The balance that apes have struck is this: When immediately valuable resources such as food and mates are at stake, competition structured by dominance rules the day. Individuals also collaborate in coalitions to obtain such resources in a zero-sum game against other individuals. Great apes, along with many other primate species, use various kinds of prosocial interactions, such as grooming and food sharing (and possibly instrumental helping), to cultivate and maintain friends—beyond kin—who may provide them with coalitionary support and other benefits. The prosocial acts in such friendly interactions are likely motivated by the prosocial emotion of sympathy, but the overall context of contest competition means that the social selection of partners is focused mainly on dominance and fighting ability.

From the point of view of cooperation, what is most exceptional about chimpanzees and bonobos, compared with other nonhuman primates, is their group hunting of small mammals. This behavior is exceptional because it occurs outside of the otherwise all-encompassing matrix of intragroup competition for a single set of zero-sum resources. Other nonhuman primates do not typically engage at all in collaborative activities aimed at producing resources unavailable to individuals. (There are a few reports of some monkey species, e.g., capuchins, mobbing or encircling small mammals, e.g., squirrels, on some opportunistic occasions [Rose et al., 2003].) We may thus consider chimpanzee and bonobo group hunting as a kind of "missing link" in the transition to human obligate collaborative foraging. Chimpanzees and bonobos do not depend for their existence on this group hunting behavior, however, and so they are not, like humans, interdependent with others more generally for acquiring their life-sustaining resources—and there is little or no partner choice or partner control in this group hunting. Other ways in which the group hunting of great apes is not human-like are elaborated in Chapter 3.

And so let us imagine the last common ancestor of great apes and humans on the model of contemporary chimpanzees and bonobos. They would have lived in complex social groups with relatively intense competition for resources, structured by social relationships of dominance and friendship, and on occasion they would have hunted in small groups for various kinds of small prey. Given other research on such things as great ape cognition and emotions, we may thus summarize the psychological prerequisites for morality that were very likely characteristic of humans' last common ancestor with other great apes as follows:

- *Cognition:* (1) skills of individual intentionality for making flexible and informed decisions, which carry with them "instrumental pressure" for choosing the best and most efficient behavioral option (instrumental rationality); and (2) skills for understanding and sometimes predicting the intentional states and decision making of others, mainly for purposes of competing with them (see Tomasello, 2014, for a review).
- *Social motivation:* (1) the capacity to form long-term social relationships of dominance and friendship with groupmates and to recognize these same social relationships among familiar third parties; (2) the capacity to have and express basic emotions in reaction to important social events, including social anger, and to recognize these emotions in others; (3) the capacity to communicate intentionally (Tomasello, 2008); and (4) the sympathy-based motivation to help others instrumentally, especially kin and friends.
- *Self-regulation:* (1) the ability to control impulses for immediate self-gratification for prudential reasons, such as avoiding conflicts with a dominant; and (2) the ability to collaborate with others to produce new resources outside the normal zero-sum resources available to individuals, which depends on, among other things, inhibiting the impulse to chase the prey individually.

Our imagined common ancestors would thus have had some sophisticated forms of cognition and sociality—including forms of cooperation and prosociality—but they would not have had any socially normative sense of fairness or justice concerning how they "ought" to treat others or how others "ought" to treat them.

We thus find ourselves in a middle theoretical position. On the one hand, theorists such as Jensen and Silk (2014; Silk, 2009) are skeptical that nonhuman primates have any prosocial sentiments or other-regarding preferences whatsoever. But we have outlined here the experimental evidence that great apes' instrumental helping is the real thing, aimed at helping others attain their goals and motivated by something like the prosocial emotion of sympathy. Jensen and Silk have interpreted the studies of instrumental helping as relying on some kind of coercion or harassment from the one needing help. But there is no compelling evidence for this view, and in the study by Greenberg et al. (2010) such coercion was specifically looked for and not found.

On the other hand, de Waal (1996, 2006) believes that many nonhuman animals, but most especially great apes, possess the roots of human morality, including not only a sense of sympathy but also a sense of reciprocity as a forerunner to concerns for fairness and justice. We have summarized evidence here supporting the first part of the proposal that great apes are sympathetic toward others, but there is no support for the second part of the proposal. Great ape reciprocity is not, we would argue, a precursor for the human notions of fairness and justice. The problem, as de Waal himself argues, is that the reciprocity involved here is only an emotion-based reciprocity. From the point of view of the interdependence hypothesis, emotional reciprocity is simply another manifestation of sympathy: friends are sympathetically motivated to help friends in various ways, and this often manifests itself bidirectionally (due to *inter*dependence). Although sympathy may be a necessary prerequisite for something like fairness—to be fair, one must be concerned to some degree about the fate of others—it is not sufficient; for fairness we need some other psychological ingredients. And our claim here is that nonhuman primates have basically none of these other psychological ingredients.

The question thus arises, what other capacities and motivations are needed to get us to something like a human sense of fairness or justice, as the second main pillar of human morality? Answering this question will require a complex evolutionary narrative involving the entire rest of the book, but we may pave the way for this account by invoking at the outset two fundamental insights of the great social theorist David Hume. The first is that there can be no fairness or justice in a social group in which there are individuals who can completely dominate and impose their will on others with impunity; for fairness, there must be some sense of equality structuring the social interactions. Hume says:

> Were there a species of creatures, intermingled with men, which, though rational, were possessed of such inferior strength, both of body and mind, that they were incapable of all resistance, and could never, upon the highest provocation, make us feel the effects of their resentment; the necessary consequence . . . is that we . . . should not, properly speaking, lie under any restraint of justice with regard to them. . . . Our intercourse with them could not be called society, which supposes a degree of equality; but absolute command on the one side, and servile obedience on the other. (1751/1957, pp. 190–191)

Substitute for "men" great apes with a high dominance status, and for "servile obedience" great apes with a subordinate dominance status, and we capture the major force in great ape social life: physical power and dominance. Dominant individuals might feel sympathy for another in need when it costs them little, but when competition over resources is at stake, why should dominants restrain their instrumental impulses for subordinates who have no leverage of any kind to make them hesitate? No, to get started down the path to fairness and justice, we need creatures who can resolve their conflicts of interest based on something other than raw power and dominance. To quote that other great social theorist of the eighteenth century, Jean-Jacques Rousseau: "Force is a physical power, and I fail to see what moral effect it can have" (1762/1968, p. 52).

Hume's second insight is that there can be no fairness or justice among individuals who are totally self-sufficient in procuring everything they need on their own; to be concerned with fairness and justice, individuals must have some sense of their dependence on one another:

> Were the human species so framed by nature as that each individual possessed within himself every faculty, requisite both for his own preservation and for the propagation of his kind . . . it seems evident, that so solitary a being would be as much incapable of justice, as of social discourse and conversation. . . . Where mutual regards and forbearance serve to no manner of purpose, they would never direct the conduct of any reasonable man. (1751/1957, pp. 191–192)

This is, of course, nothing other than a terse statement of our principle of interdependence from a negative point of view: individuals who have no need of others have no need for fairness and justice. Chimpanzee and bonobo individuals are dependent on others in some ways, but they do not depend on others to obtain the basic resources they need to survive and thrive, and they very likely have no understanding that each is mutually dependent on the other. A key task in our natural history to come, therefore, will be to establish how early human individuals became interdependent with one another in some new and especially urgent ways, and also became aware of this interdependence so that it figured in their rational decision making. The goal is to show how humans' interdependent interactions with one another—specifically, the proximate psychological mechanisms underlying these interdependent interactions, including a sense of "we" as plural agent—prompted early humans to construct a new and species-unique sense of fairness and justice.

And so we may hypothesize that the last common ancestors of humans and great apes were at least somewhat prosocial creatures, that is, toward kin and friends and in the overarching context of intragroup competition. Though modest, this starting point cannot be ignored because, in point of fact, much of human morality, in a very broad sense, is based on this kind of sympathy for particular others, including especially friends and family. Humans have not left this moral dimension behind; they have simply developed some other forms of morality on top of this that have led them to care for and respect a wide variety of other human beings with whom they are less intimate—not only because they sympathize with them but also because they feel they ought to.

3

Second-Personal Morality

> The standing to make claims and demands on one another as free and rational persons is something to which we are jointly committed whenever we take up the second-person stance.
>
> —STEPHEN DARWALL, *THE SECOND-PERSON STANDPOINT*

Despite the fact that they live in cooperative social groups and behave prosocially toward kin and friends—and are of course worthy targets of our moral concern—chimpanzees themselves are not moral agents. We do not allow them to roam freely in our midst for fear they will attack our children, steal our food, destroy our property, and generally wreak havoc without regard for anyone else. And if they did all of these antisocial things, no one would blame them or hold them responsible. Why not? Why are they not moral agents? The data presented in Chapter 2 suggest that the answers to these questions are not simple things, like they do not act with instrumental rationality, they do not comprehend the goals and desires of others, they do not feel emotions or comprehend the emotional expressions of others, they do not engage in any types of prosociality, or they do not control their impulses when they need to do so. They do all of these things perfectly well in at least some situations.

No, the answers to these questions—the many and various reasons that chimpanzees are not moral agents—are embedded in the whole complicated story we are about to tell about how human beings gradually became ultrasocial, ultracooperative, and, in the end, moral apes. The animating idea is that from a great ape starting point, human beings began down a moral path as they adapted to some new forms of social life, specifically, in the beginning, some new forms of interdependent collaborative activity. Following the lead of social contract theorists from Hobbes to Rawls, then, we suppose that the natural home of human morality—with special reference to issues of fairness and justice—is cooperative activity for mutual benefit. "The primal scene of morality" says Korsgaard (1996b, p. 275) "is not one in which I do something to you or you do something to me, but one in which we do something together."

Our argument is that participation in certain kinds of mutualistic collaborative activities selected for individuals who were able to act together dyadically as a joint agent "we." This required both partners to have the skills and motivations necessary for creating a joint goal guided by joint attention, which then structured their individual roles and perspectives (i.e., joint intentionality; Tomasello, 2014). Evolutionarily, the problem was that the group hunting of humans' last common ancestor with other apes (using contemporary chimpanzees as a model) did not create enough or the right kind of interdependence (individuals could opt out and still do fine) or enough or the right kind of partner choice (individuals could participate or not whenever they pleased) to force the issue. What was needed to make individuals recognize and trust their interdependence with a partner, and so to form a joint intentional "we," was collaborative activities on which everyone depended absolutely, no escape, and a system of partner choice and control that made everyone accountable to everyone else for treating collaborative partners with the respect they deserved. What was needed as context for the emergence of joint intentionality was *obligate* collaborative foraging with various and robust means of partner choice and control.

Our proposal will be that, initially, early humans collaborated in joint intentional activities for purely strategic reasons, using others as a kind of "social tool" to further their own interests. Their collaboration was thus what social theorists would call contractarian. But over time, we will argue, interdependent collaborative activities structured by joint intentionality fostered in participants a new kind of cooperative rationality. They came to understand that particular collaborative activities had role ideals—socially normative standards—that applied to either of them indifferently, which implied a kind of self–other equivalence (see Nagel, 1970). Based on the recognition of self–other equivalence, there arose a mutual respect between partners, and a sense of the mutual deservingness of partners, thus creating second-personal agents (see Darwall, 2006). Such second-personal agents had the standing to make joint commitments to collaborate and to jointly self-regulate their collaboration (see Gilbert, 2014). The joint intentional activity was now what social theorists would call contractualist, since it was constituted by an actual agreement between second-personal agents. The outcome was what we may call a second-personal morality: a dyadic morality of face-to-face interactions between the second-personal agents "I" and "you" (perspectivally defined) collaborating together, and feeling responsible to one another, as a jointly committed "we." This new morality existed only within the collaborative engagement itself, or when considering such an engagement, and not in other domains of life. Outside of these special collaborative contexts, the social interactions of early humans would have remained almost totally ape-like.

The procedure in this chapter is as follows. We first look at early humans' new forms of interdependent collaboration and how these gave rise to an expansion in sympathetic concern and helping, the first morality of sympathy. To characterize the first morality of fairness, we then look at three sets of psychological processes. First are the cognitive processes of joint intentionality that structured early humans' collaborative activities via a new form of plural ("we") agent. Second are the social-interactive processes of second-personal agency that arose in the context of early humans' partner choice and control, creating in participants a sense of mutual respect and deservingness. Third are the self-regulatory processes set into motion by joint commitments, designed to keep the collaboration on track until its consummation, a joint self-regulation that created for each partner a number of responsibilities to the other.

Methodologically, although there are no beings alive today identical to the earliest moralists we are envisioning here, we may nevertheless cite as relevant various lines of empirical research with young children, mostly at or under three years of age. Two- and three-year-old children are at least somewhat appropriate analogues, we would argue, because they engage directly with others in dyadic interactions and relationships—their "interaction engine" for face-to-face encounters has started (Levinson, 2006)—but they seemingly have few if any social skills for operating in groups qua groups. Children of this age thus have some species-unique skills and motivations for moral action and judgment toward individuals, but they are still not participating actively in the social conventions, norms, and institutions of their cultural group.

Collaboration and Helping

We begin with the origins of early humans' more expansive (compared with other great apes') helping and sense of sympathetic concern for others. That is to say, we are concerned in this section with how great apes' prosociality toward kin and friends became early humans' morality of sympathy. It arose, we will argue, in the context of a general "taming" of the species and a newly collaborative way of obtaining basic resources.

Self-Domestication

The initial move in the direction of human morality was addition by subtraction. Specifically, what had to be subtracted was great apes' almost total reliance on dominance—either by individuals or by coalitions—to settle any and all disputes. Individuals had to become less aggressive and less bullying if they were going to forage together collaboratively and share the spoils peaceably at the end. The proposal is that this began happening soon after the emergence of the genus *Homo* some 2 million years ago in a transformation that may be thought of as a kind of self-domestication (Leach, 2003; Hare et al., 2012).

This transformation was very likely the product of three intertwined processes. First, early humans began mating via pair bonding. This new form of mating (i.e., new among great apes) had many cascading effects on human emotions and motivations. Chapais (2008) points out that pair bonding also produced, via the same mechanism as sibling recognition, a new recognition of paternity in both directions: everyone who hangs around the female to whom I am bonded is my close relative. While this new form of bonding had many ramifications, most important for current purposes is the fact that males now recognized all of their offspring (as well as siblings and mating partners) in the social group, and this would have led them to be less indiscriminately aggressive.

A second part of the story involved a new subsistence strategy. Most theorists who emphasize the importance of collaborative hunting of large game in human evolution recognize a transitional stage of scavenging. Individuals would have been forced to work together in a coalition to chase away the lions or hyenas feasting on a carcass before they themselves could scavenge (Bickerton and Szathmáry, 2011). Any individual who then hogged all the meat would have been the target of another coalition aimed at stopping him. Boehm (2012) has emphasized that, in general, almost all contemporary hunter-gatherer groups are highly egalitarian, and overly dominant individuals are quickly brought down to size by coalitions of others. Evolutionarily this would have meant that there was social selection against bullies, food hogs, and other dominants, and thus social selection for individuals who had a greater tolerance for others in cofeeding situations. Indeed, in modern-day chimpanzees, collaboration in an experimentally created foraging task goes best when the pair is made up of individuals who are tolerant of one another around food (Melis et al., 2006b).

Third, although we do not know precisely when it evolved, humans are the only great ape that practices collaborative childcare (also known as cooperative breeding; Hrdy, 2009). Individuals who are not parents, and sometimes not even close kin, nevertheless help to feed and care for children. While fathers and grandparents would have had obvious evolutionary bases for their helping, nonkin might have evolved a tendency to care for others' children in the context of collaborative foraging. Among contemporary hunter-gatherers, women can be much more productive in gathering plant foods if they do not have several children to care for at the same time. We might thus speculate that cooperative childcare evolved in tandem with collaborative foraging, as a division of labor to maximize food production, assuming that the gatherer would share her bounty with the caregivers. Cooperative breeding nonhuman primates, such as marmosets and tamarins, seem to have a number of more prosocial tendencies compared with other monkeys, and some theorists have even speculated that cooperative breeding by itself is the main instigator of the great amount of prosocial behavior in modern humans (e.g., Hrdy, 2009; Burkart and van Schaik, 2010).

The outcome of these three precursor processes was the evolution of less dominance-based social interactions and more gentle personal temperaments, a move toward Hume's first prerequisite of a greater balance of power among individuals (see Chapter 2). It was thus this pair-bonded, child-caring, relatively tolerant and gentle creature—a self-domesticated great ape—who entered into the new and still more collaborative lifeways that we will be positing as the evolutionary origins of uniquely human cooperation and morality.

Obligate Collaborative Foraging

Now came the truly crucial step. As always in evolution, the instigation was an ecological change. About 2 million years ago, as the genus *Homo* was emerging in Africa, a global cooling and drying spell created an expansion of open environments and a radiation of terrestrial monkeys (e.g., baboons), who may have outcompeted *Homo* for many resources. This meant that early humans' preferred foods (i.e., fruit and other nutritious vegetation) became scarce, and new options were needed. Scavenging large carcasses killed by other animals would have been one such option that coalitions of individuals could exploit.

But at some point early humans began more active attempts at hunting large game. A good guess would be that this tendency, though begun earlier, was fully consolidated by the time of *Homo heidelbergensis* (the common ancestor to Neanderthals and modern humans) some four hundred thousand years ago. There is much evidence, for example, that these creatures hunted large game collaboratively and systematically (see Stiner, 2013, for a review). The payoff structure in this and other forms of collaborative foraging would likely have been that of a stag hunt: individuals could opt for an easy, low-value food or, by collaborating with a partner, opt for a riskier but higher-value alternative (Skyrms, 2004). Whereas other apes obtained (and still obtain) the vast majority of their nutrition mainly through solitary efforts, these early humans obtained the vast majority of their food mainly through collaborative efforts. And, importantly, they had no or few satisfactory fallback options if the collaboration failed. The collaboration was obligate. This meant that individuals were interdependent with one another in much more urgent and pervasive ways than were other apes: they had to collaborate with others on a daily basis or else starve.[1]

For collaborative foraging to get off the ground, of course, each partner must expect some benefit. Thus, chimpanzee group hunting of monkeys works because each individual hopes it will catch the monkey itself and can count on getting some scraps even if it does not. But in situations in which the spoils can be monopolized by the captor, things fall apart. For example, in an experimental study Melis et al. (2006b) presented pairs of chimpanzees with out-of-reach food that could be obtained only if both partners pulled simultaneously on the two ends of a rope connected to a platform. When there were two piles of food, one in front of each individual, the pairs were often successful. However, when there was only one pile of food in the middle of the platform, pulling it in typically resulted in the dominant individual monopolizing all the food. This naturally demotivated the subordinate for future collaborative efforts, so cooperation fell apart over trials. In stark contrast, in a study designed by Warneken et al. (2011) to be as comparable as possible to this one, three-year-old children were not bothered at all by the food being in a single pile in the middle of the board; they collaborated successfully over many trials no matter how the food was laid out. The children knew that, no matter what, in the end they could always work out a mutually satisfactory division, which they almost always did (see Ulber et al., submitted, for something similar with even younger children). We may thus imagine that when early humans were forced to forage collaboratively, those who were most successful over time were those who naturally shared the spoils among themselves in a mutually satisfactory manner.

Needless to say, the collaboration could turn out badly if individuals chose either incompetent or greedy partners. Early humans' foraging interdependence was thus supplemented with a robust system of partner choice. As noted in Chapter 2, in chimpanzees' and bonobos' group hunting of monkeys there is no good evidence for partner choice. But early humans experienced much stronger pressures to selectively seek good partners and avoid poor ones—again due to a paucity of fallback options—so social selection for good collaborators gradually emerged. Only individuals who could work well with others ate well and so passed on their genes prolifically. And, as we shall see later, early humans also evolved very strong means of partner control, by which they attempted to turn poor partners into good ones.

These seemingly small changes from great ape group hunting to early human collaborative foraging—as the collaboration became ever more obligate and the partner choice and control assumed ever greater importance—completely restructured the social-ecological demands on individuals. Various novel adaptations to these demands are explored in the other sections of this chapter; the remainder of this section focuses on the most basic: a quantitative increase, and perhaps a qualitative change, in the sympathetic concern that individuals showed for one another.

Concern for Partner Welfare

Many people have speculated that sympathy for others originated in concern for one's offspring. Chimpanzees and other great apes, as previously documented, may in addition feel sympathy for, and so help, friends on whom they are dependent for coalitionary and other forms of support. Our proposal here is simply that as humans became more deeply interdependent with a wider variety of individuals in the context of obligate collaborative foraging, a deeper and wider application of sympathetic concern and helping emerged.

The main point is this: in the context of obligate collaborative foraging, helping one's partner pays direct dividends. If my partner has dropped or broken his spear, it is in my interest to help him find or repair it, as this will improve our chances for joint success. In this context the helped partner has no incentive to now suddenly defect and run away, as the mutualistic situation that will benefit him is still operative, so he takes the help and continues the mutualistic collaboration. It is thus not surprising that when Ache foragers in South America are hunting, they do such things for their partners as give them weapons, clear trails for them, share information with them, carry their children for them, repair their weapon for them, and instruct them in best techniques (Hill, 2002). There are no reports of chimpanzees helping one another in their group hunting of monkeys, presumably because they are mainly competing to capture the monkey, although in experiments in which the food competition is eliminated even they will help their partner (Melis and Tomasello, 2013). Rational agents feel instrumental pressure to act toward their goals (desires) given their perceptions (beliefs), so each partner in a collaboration feels instrumentally rational pressure to help her partner as needed to further their joint enterprise.

Unlike the simple cases imagined by most modelers of human evolution, what we are imagining here are situations in which human behavior is hierarchically organized such that some actions are performed to meet goals at multiple levels. Helping my partner may be a hindrance with respect to my subgoal of flushing out the prey—because stopping to help him delays me—but in the larger context of the mutualistic activity as my overarching goal, stopping to help my partner is directly beneficial. And so in early humans sympathetic concern and helping extended beyond kin and friends to collaborative partners in general, independent of any personal relatedness or personal history of cooperation. And for individuals who had any sense of the future, the logic of interdependence demanded that they also help potential partners outside of the collaborative activity itself because they might be needing them in the future. If the partner with whom I have daily success is hungry tonight, I should feed her to keep her in good shape for tomorrow's outing. It is noteworthy, to repeat, that this account does not depend on reciprocity in the classic sense because the individual is "repaid" for his altruistic act not by costly reciprocated altruistic acts from the beneficiary but, rather, by her mutualistic collaboration with him, which costs her nothing (actually benefits him) and which she would do in any case.

Although interdependence is a part of the evolutionary logic of human altruism, it need play no role in the individual's personal decision making; the proximate motivation may simply be to help anyone with certain characteristics or within a certain context. Indeed, evidence for this possibility may be found in recent experimental research establishing that young children are intrinsically motivated (perhaps by sympathetic concern) to help others:

- From a very early age, humans are highly motivated to help others. Infants as young as fourteen months will help unfamiliar adults with various problems, such as fetching out-of-reach objects, opening doors, and stacking books (Warneken and Tomasello, 2006, 2007). Two-year-olds will help others in such ways even at a high cost to themselves (Svetlova et al., 2010).
- From a very early age, humans are internally motivated to help others, with no need for external incentives. Two-year-olds do not help more when their mothers are watching or encouraging them to help (Warneken and Tomasello, 2013); they help someone who does not even know he is being helped (Warneken, 2013), and when they are given external rewards for helping, if the rewards are then stopped their helping actually decreases relative to children who were never rewarded (Warneken and Tomasello, 2008).
- From a very early age, human helping is mediated by a sympathetic concern for the plight of others. Infants go to some trouble to comfort another person showing signs of emotional distress (Nichols et al., 2009), and indeed, the more they are distressed about the others' situation, the more likely they are to help (Vaish et al., 2009).

In all, it would seem that helping others comes naturally to young humans, and it is intrinsically motivated by feelings of sympathy.

Even though the proximate motivation for human helping does not seem to include calculating one's dependence on the help's recipient, there are two important pieces of evidence suggesting that interdependence is nevertheless a key part of its evolutionary foundation. First, using a direct physiological measure of emotional arousal, pupil dilation, Hepach et al. (2012) found that young children are equally satisfied both when they help someone in need and when they see that person being helped by a third party—and more satisfied in both of these cases than when the person is not being helped at all. This suggests that their motivation is not to provide help themselves but only to see that the other person is helped. Importantly, this means that a concern for reciprocity (either direct or indirect) cannot be the evolutionary basis for young children's helping because to benefit from reciprocity one has to perform the act oneself so that one can be identified as the helper and later paid back. But if the proximate motivation is just that the other person be helped, this would fit very nicely with an evolutionary account in terms of interdependence in which I do not care how the person is helped as long as he is helped; I only care about his well-being.

Second, it is an interesting and important fact that humans are especially motivated to help others who are in dire physical straits, for example, the proverbial stranger lying injured on the street. We are urgently concerned for his physical well-being, so much so that even if he tells us that we should first retrieve his bicycle from the street before we try to stop his bleeding, we ignore his wishes and try to stop the bleeding anyway. We are concerned not as much with fulfilling his wishes as with helping him physically (see Nagel, 1986). Several studies of young children find such paternalistic helping, as it is called, from a fairly young age (e.g., Martin and Olson, 2013). The point is that we do not help others to get what they want if this conflicts with what they need from the point of view of their physical well-being. This view is further supported by the finding of Hepach et al. (2013) that young children do not help another when her expressed need seems unjustified by the circumstances (e.g., she is crying about something trivial), but only when her expressed need seems justified (see Smith, 1759/1982). This finding suggests, again, that it is not just the personal desires of others that children and adults are attempting to accommodate—although they may do this in some cases; rather, they are mainly looking out for the well-being of the help's recipient. Paternalistic helping, selectively dispensed only to those who really need it, is of course consistent with an evolutionary logic of attempting to keep potential collaborative partners in good shape.

Important for our overall account as well are two further findings. First, when young children are given the opportunity to help another child, they do so more readily if this takes place in the context of a collaborative activity than in some neutral context (whereas chimpanzees do not show this effect; Hamann et al., 2012; Greenberg et al., 2010). This provides fairly specific support for an evolutionary scenario in which the primordial context for the expansion of human altruism toward nonfriends is mutualistic collaboration.

The second finding is that human infants, but not great apes, will help another more when that other has previously been "harmed" by a third party than when he has not (Vaish et al., 2009; Liebal et al., 2014). This suggests the possibility of a qualitatively different form of sympathy, one that goes beyond merely helping an agent with her instrumental problem to actually empathizing with her, in the sense of taking her perspective and putting oneself "in her shoes." Roughley (2015) calls this "Smithian empathy" and gives as examples (following Smith, 1759) the way in which we feel bad for a dead or mentally incapacitated person—even though they themselves are not feeling bad—based presumably on how I would feel if I (in my current aware state) were in her situation. The paternalistic helping study cited above may also be interpreted in this way, of course: I fetch him not what he wants but what I would want if I, given my current knowledge, were in his shoes. This perspective taking and self-projection are presumably things that apes do not do, and they rest, as we argue below, on the uniquely human sense of self–other equivalence.

And so the first component of our first step on the road to modern human morality is an expanded sympathetic concern for nonkin and nonfriends, which leads to helping them and, possibly, to a qualitatively new Smithian empathy in which the individual identifies with another in his situation based on a sense of self–other equivalence. Because of interdependence, this sympathy and empathy for others presumably contributes to the helper's reproductive fitness on the evolutionary level, but, to repeat, the evolved proximate mechanism contains nothing about interdependence and reproductive fitness. It is based only on an unalloyed sympathy for others, which may then compete with a variety of other motives, including selfish motives, in the actual making of behavioral decisions.

Nevertheless, however pure and powerful it may be, concern for the welfare of others is not, by itself, sufficient to account for the other major pillar of human morality: the morality of fairness based on a sense of obligation. The morality of fairness involves not helping others but, rather, balancing their (and one's own) many and varied, sometimes conflicting, concerns as they interact in complex situations. To account for this novel moral attitude, we will need to specify three additional sets of psychological processes that also emerged in the context of early humans' obligate collaborative foraging:

- cognitive processes of joint intentionality,
- social-interactive processes of second-personal agency, and
- self-regulatory processes of joint commitment.

We do this, each in turn, in the three sections that follow.

Joint Intentionality

In his 1871 book *The Descent of Man,* Darwin writes: "Any animal whatever, endowed with well-marked social instincts, the parental and filial affections being here included, would inevitably acquire a moral sense or conscience, as soon as its intellectual powers had become as well, or nearly as well developed, as in man" (p. 176). In the absence of detailed comparative research on the cognitive skills of humans and other animals, Darwin could not know precisely which uniquely human "intellectual powers" were the crucial ones. But the current proposal—backed by the systematic comparative research reported here—is that now we do know, at least in general outline. The uniquely human cognitive abilities responsible for the evolution of human morality, especially with regard to a sense of fairness and justice, all fall under the general rubric of *shared intentionality* or, more precisely, at this first step in our story, *joint intentionality* as effected by a collaborating dyad. Not themselves moral (or fair or just), these abilities nevertheless provided the necessary foundation for a key first step in this direction.

The Dual-Level Structure of Joint Agency

A collaborative activity structured by joint intentionality possesses a dual-level structuring of jointness and individuality: each individual is both the "we" that is pursuing with her partner a joint goal (in joint attention) and at the same time an individual that has her own role and perspective (Tomasello,

2014). Participating in such joint intentional activities connects the partners with one another psychologically in a unique way: they now form what we may call a joint agent. Thus, if in the face of a rainstorm two people each run to the same shelter, the seeking of shelter is not their joint goal but only the individual goal of each of them separately (Searle, 1995). Chimpanzees hunting monkeys are like this; they do not have a joint goal of capturing the monkey; rather, each has the personal goal of capturing the monkey for itself (as evidenced by their attempts to monopolize the carcass afterward). In contrast, a joint agent is created when two individuals each intend that "we" act together jointly toward a single end, and they both know together in common ground (they both know that they both know) that this is what they both intend (see Bratman, 1992, 2014).

The formation of a joint goal relies on a mutual sense of trust—at the very least, a mutual sense of "strategic trust" that we both know, and know together that we know, what it is individually rational for each of us to do for joint success. (Below we introduce the richer notion of "normative trust" that binds partners together into a still stronger "we" on the basis of their explicit joint commitment.) Young human children are able to create with others joint goals soon after their first birthdays. Thus, when fourteen- to eighteen-month-olds are collaborating with an adult partner who just stops interacting for no reason, they make active attempts to reengage him in the task by doing such things as beckoning and pointing—whereas chimpanzees in this same experimental situation never attempt to reengage their partner at all (Warneken et al., 2006). And the children in this experiment are not just attempting to reactivate the adult as a "social tool" toward some personal end: when there is some external reason for the interruption (e.g., the adult is called away by another adult), children wait patiently for his return, *even if they could easily perform the task by themselves* (whereas they continue attempting to reengage him if he stops cooperating for no discernable reason; Warneken et al., 2011). The reengagement attempts are thus sensitive to the adult's intentional state—if he is called away he likely retains the joint goal, but if he quits for no reason he has likely lost it—and children are attempting not just to reinstate a fun activity but, rather, to reconstitute their lost "we."

Epistemically, when two individuals act together jointly, they naturally attend together jointly to situations that are relevant to their joint goal. Joint intentional activities thus have as a crucial component joint attention between participants. Analogous to the rain shelter example above, two individuals both attending to the same situation at the same time is not sufficient for joint attention; they must attend to it together, in the sense that they both know that they are attending to it together. Defined in this way, great apes do not engage in joint attention (see Tomasello and Carpenter, 2005, for empirical evidence), whereas human infants engage with others in joint attentional episodes from about nine to twelve months of age, with these interactions forming the basis for everything from collaborative activities to intentional communication and language acquisition (Tomasello, 1995). When two individuals have experienced things together in joint attention, this shared experience becomes part of their personal common ground, for example, that we both know together what needs to be done to gather honey. Personal common ground is part of what defines a social relationship, and individuals rely extensively on their personal common ground with others to make many crucial social decisions (Moll et al., 2008; Tomasello, 2008).

Individuals participating as part of a "we" in a joint intentional activity do not thereby lose their individuality. On the behavioral level, each has her own individual role to play, and in the normal case, they each know both roles. Thus, in a recent study three-year-old children first played one role in a collaborative activity and then were forced into the other role. Their performance showed that they had learned a lot about this other role already, since they were much more skillful at it than were naïve children. This "value added" from playing the opposite role did not hold for chimpanzees, however, suggesting that they were participating with their partner in a more individualistic fashion (Fletcher et al., 2012). Importantly, as they simulate the partner in her role, even very young children understand that they can reverse roles with her and still achieve joint success (Carpenter et al., 2005), and again, this is not true of chimpanzees (Tomasello and Carpenter, 2005). An understanding of role interchangeability suggests that the participating individuals conceptualize the collaborative activity as a whole from a "bird's eye view," with both the self's and the partner's perspective and role in the same representational format.

On the epistemic level, each partner in an act of joint attention also has his own individual perspective and knows something of his partner's perspective as well. Indeed, one could argue that joint attentional engagement is what creates the whole notion of perspective in the first place: we need to be focused on something in common for the notion of different perspective on it to even arise (Moll and Tomasello, 2007a). Strong evidence for this interpretation comes from the fact that infants do not register what others are experiencing when they simply watch them acting from afar; they only take the other's perspective into account when they are interacting together with him jointly (Moll and Tomasello, 2007b). Such perspective taking is presumably what enables each partner in a joint activity to monitor the role of the other (and also the other monitoring him).

To coordinate skillfully the differing perspectives, and thus roles, in a joint intentional activity, some kind of communication is needed. Tomasello (2008) argues and presents evidence that, indeed, joint intentional activities were the birthplace of humans' unique forms of cooperative communication, beginning with the natural gestures of pointing and pantomiming. Cooperative communication is distinguished from great ape communication in general by the fact that the communicator is informing the recipient of something that will be useful or helpful to her, and the recipient, because the communication occurs within the context of a mutualistic collaborative activity, trusts the information. Importantly in the current context, when something is explicitly communicated it is "out in the open" in the joint attention and common ground of the interactants. This means that there can be no question of not knowing this information, and so if it has interpersonal consequences one cannot hide from them by claiming ignorance. Thus, if I ask for your help explicitly, using communicative means that we both know in common ground you comprehend, you cannot act as if you do not know I need help. This "openness" will play a critical role later in our account in the creation of joint commitments, which create interpersonal feelings of "ought" in the collaborative act (normative trust and second-personal responsibility). In general, the common-ground understanding of a situation is an integral part of any normative dimensions it may have, and overt communication is the strongest way to create such common-ground understanding.

A crucial outcome of all of this was that the dual-level cognitive structure of joint intentionality created for early humans a new type of social relationship beyond friendship: the relationship that "I," as collaborative partner, have with "you," as my collaborative alter ego, in the context of our "we" as joint agent that is working, within the context of our personal common ground, toward a joint goal (with, of course, "I" and "you" perspectivally defined). And so was born the new cast of characters about whom individuals must care if they hope to be successful in acts of joint intentionality: "I," "you," and "we," the essential ménage à trois from which the human morality of fairness has sprung.

Collaborative Role Ideals

As early human partners collaborated with one another repeatedly, there arose between them a common-ground understanding of the ideal way that each role should be played in a particular foraging activity. For instance, as a dyad interacted repeatedly in, say, antelope hunting, they developed a common-ground understanding of what constituted the ideal performance of, say, the chaser role. Because these ideals existed only within a particular dyad's common ground with respect to a particular collaborative activity, we may call them role-specific ideals: our subgoal for whoever is playing this role. And of course, through some kind of abstraction there could develop more general ideals that apply to all roles alike, such as exerting effort, sharing the spoils, and not abandoning the collaboration for trivial distractions. As a part of this common ground, partners of course knew that if either of them did not perform his role in the ideal way, there would be joint failure. Thus, there was both positive instrumental pressure on each partner toward success and negative instrumental pressure on each partner against failure, with these differing outcomes affecting them both more or less equally. And this meant that there were also effects on their plural agent "we," in the sense of their continuing partnership in general.

We may think of this common-ground understanding of ideal role performance—virtuous performance, if you will—as constituting the strategic roots of socially shared normative standards. Of course, these primitive role ideals are not yet the kind of normative standards to which moral philosophers ascribe such overriding importance; they are still fundamentally instrumental and local. But neither are they a simple extension of individual instrumentality. They are, rather, a kind of socialization of individual instrumentality, as may be seen in three critical features that go well beyond anything in the individual intentionality of other great apes. First, the role ideal is not simply the goal of an individual agent; rather, it is the subgoal of the joint agent "we," since it makes sense only in the context of a joint goal with coordinated roles (e.g., chasing an antelope makes sense only if there is someone waiting ahead to spear it). Second, it is in the common-ground knowledge of the partners that success or failure in meeting a role ideal affects not only the individual playing that role but also the valued partner ("you"), as well as the more long-term partnership ("we"), which adds a social dimension to the instrumental pressure on each partner. And third, the role ideal specifies what an agent in a role should do to contribute to the joint agent's success, but it is at the same time partner independent: it applies to anyone and everyone alike irrespective of any personal characteristics and/or social relationships. We may thus say, in general, that the individual instrumental rationality of great apes has become socialized into the joint instrumental rationality of a pair of early humans acting as a joint agent.

In all, it is difficult to imagine some origin for socially shared normative standards other than common-ground role ideals in collaborative activities. The main alternative is that the individual somehow feels pressured by a multitude of others (or one especially powerful other) to behave in a certain way (e.g., von Rohr et al., 2011). But then the standard is "their" standard, and my adherence to it is purely a matter of individual instrumentality or strategy with respect to an external constraint. In contrast, the common-ground role ideals of joint intentional activities transcend this individuality because they are socially shared standards that the role performer himself endorses (see below on joint commitment), and the upholding of which facilitates success not only for the individual himself but also for his valued partner and partnership.

Self–Other Equivalence

How the partners in a joint intentional activity understand what they are doing is crucial. In a joint intentional activity, we have claimed, the individual understands what she is doing from a kind of "bird's eye view." She does not peer from inside her role and perspective onto the outside of the partner and what he is doing. Rather, as she is collaborating the individual imagines being in the partner's role and perspective, on the one hand, and also imagines how the partner is imagining her role and perspective, on the other. She can imagine either person in either role (aka role reversal) because, as we have just argued, she understands that the instrumental demands of the roles are partner independent; in getting the job done, all persons are equivalent. (This is not to say that there are no individual differences in how well a role is performed, only that success results from the individual, whoever that may be, performing the role in its ideal way.) This bird's eye view of the collaborative activity is thus, in an important sense, impartial. It does not matter if you are my friend, a child, my mate, or an expert, it is the partner-independent roles and their successful execution that matters, not any personal characteristics of those who are playing them.

The effects of this way of viewing things on the interpersonal relations between individuals were momentous. Nagel (1970) argues that the recognition of other persons as agents or persons just as real as oneself—such that the self is seen as just one agent or person among many—provides a reason for considering their concerns as equivalent to one's own. His description of what we have called the "bird's eye view" and the reversibility or exchangeability of roles is this: "You see the present situation as a specimen of a more general scheme, in which the characters can be exchanged" (p. 83). This forms the basis for what Nagel considers the most basic argument a victim can present to a perpetrator: "How would you like it if someone did that to you?" (i.e., if roles were reversed; p. 82). Our claim here is simply that a recognition of self–other equivalence arose in human evolution as an insight into the instrumental logic of collaborative activities with the dual-level structure of joint goals and individual roles with associated role ideals.

But crucially—and this is Nagel's deepest philosophical point—the recognition of self–other equivalence is *not* by itself a moral notion or motivation; it is simply the recognition of an inescapable fact that characterizes the human condition. I might ignore this insight in my actual behavioral decision making, and indeed, I might even wish it were not true. It does not matter; a fact is a fact. The recognition of self–other equivalence is thus not in any way sufficient for making a fair or just decision in one's interpersonal relations with others; it is simply the structure of the way that humans understand the social world in which they live. But even though it is not sufficient to motivate acting with fairness and justice, viewing oneself and others as in some sense equivalent is, as everyone since Aristotle has recognized, necessary.

In the current account, then, the recognition of self–other equivalence does not by itself constitute any kind of moral action or judgment; rather, it serves to transform various actual actions and judgments from strategic or contractarian bases—sharing food with others only because of predicted future benefits for myself—to more moral or contractualist bases in which individuals make decisions and judgments at least to some degree impartially, with self and other on a fundamentally equal plane. The recognition of self–other equivalence had its most direct effect on the new ways that early human individuals treated, and expected to be treated by, their collaborative partners. In particular, in the context of partner choice, it led collaborators, or potential collaborators, to treat one another with mutual respect as equally deserving partners. These new ways of treating partners and potential partners constituted what we may call second-personal agency (see the next section), which endowed early human individuals with the tools to create with one another actual social contracts in the form of joint commitments (see the section after next).

Summary

Early humans created within their cooperatively interacting dyads a new social order that existed on two levels simultaneously: on one level was a "we" created from the mutual recognition of partner interdependence (based on strategic trust) that could act as a joint agent; on the other level were two "I"s constituting this "we" (each perspectivizing the other as "you"), who mutually recognized their equivalence in the collaborative activity as agents under instrumental pressure to conform to a role-specific ideal. The joint agent so constituted thus had its own novel form of instrumental rationality that motivated each partner to help and share with the other. Once early human individuals began conceptualizing themselves and their equal partner as a joint agent "we"—and worrying about how "we" related both to "you" and to "me"—the basic problems for which a morality of fairness was the solution were set.

Second-Personal Agency

Early humans' joint intentional activities were thus instrumentally rational in newly social ways. There was nothing specifically moral about these activities, beyond sympathy, but the common-ground understanding of partner-independent role ideals (as a precursor to socially shared normative standards) and self–other equivalence (as a precursor to impartiality) were cooperative seeds that would soon bear moral fruit. For this moral fruit to actualize, however, individuals were needed who were more thoroughly adapted for obligate collaborative foraging, specifically, for the challenges of partner choice within a larger pool of potential collaborators. Most vital, early human individuals had to learn to create beneficial partnerships with others by evaluating and so choosing good collaborative partners, by anticipating others' evaluations and thus acting so as to be chosen as partners themselves, and, in general, by managing and controlling their ongoing partnerships in satisfactory directions.

Partner Choice and Mutual Respect

Partner choice means selecting as a collaborative partner one individual over others, presumably the one who is most competent (e.g., knowledgeable and physically adept) and cooperative (e.g., inclined to do his share of the work and to take only his share of the spoils). In the context of obligate collaborative foraging, not being chosen by any partners would of course be fatal.

In a marketplace of partner choice, the basic "positive" skill is identifying good cooperative partners. Interestingly, both great apes and very young human infants already have preferences for cooperative individuals. For example, in a pair of recent studies, chimpanzees and orangutans chose to approach and beg food from a human who had performed prosocial acts—either toward them or toward a third party—rather than from a human who had performed antisocial acts (Herrmann et al., 2013). In the same vein, human infants show similar social preferences. For example, infants one year old and younger already prefer to interact with people who are "helpers" versus "hinderers" (e.g., Kuhlmeier et al., 2003; Hamlin et al., 2007). The "negative" version of this same process is avoiding poor partners. Chimpanzees systematically avoid partners with whom they have previously failed in collaborative tasks (Melis et al., 2006a), and young human children selectively withhold help or resources from an individual whom they perceive to be somehow "meaner" than another (Vaish et al., 2010).

Early humans thus formed evaluative attitudes about the cooperative behavior of other individuals. But, unique among primates, early humans also knew that others were forming evaluative attitudes about them—and so they tried to influence the process. Thus, in a recent experiment, five-year-old human children were given the opportunity to either help or steal from another fictitious child. In some cases they did this while being watched by a peer, and in other cases they did it while in the room alone. As might be expected, they helped the other child more, and stole from her less, when they were being watched by a peer than when they were alone. Chimpanzees in the same situations did not care whether they were being watched or not (Engelmann et al., 2012). In another experimental situation designed to be similar in structure to the situation faced by early humans, individuals actually competed with one another to be more helpful and generous so that an observer would see them as better cooperators and so choose them as partners for a mutualistic collaboration (Sylwester and Roberts, 2010). The early humans we are picturing here in the process of partner choice, then, not only formed evaluative attitudes about others but also knew that others were doing the same with respect to them, and so they engaged in an active and strategic process of impression management (Goffman, 1959).[2]

The result was that early humans would have begun keeping track of the specific cooperative and noncooperative acts that they had experienced, as both actor and recipient, with specific partners. And, of course, everyone preferred to work with the most competent and cooperative partners. This opened the door to some individuals being more in demand as partners and then using this popularity as bargaining power to get a better deal. As outlined most systematically by Baumard et al. (2013), in a completely open marketplace of partner choice (with full information), the most competent and cooperative partners—those with whom one is most likely to be most successful—would be most in demand. They could therefore require of a prospective partner more than half the spoils, or the easiest role, or whatever, because the "weaker" partner would want the "stronger" one to collaborate with him again in the future more than the reverse. In this view, the marketplace was highly competitive.

But there are other kinds of marketplaces under different conditions. For example, competition will not be so keen if there are lots of good foraging alternatives that do not require exceptional partners—anyone will do—or if collaborative foraging most often begins opportunistically with whoever is nearby, severely restricting the available choices of partner. But most important in the current context, a totally open marketplace would not be operative if information about partners were limited to direct personal experiences with different partners, with nothing in the way of public reputation based on gossip; then it would be extremely difficult for single individuals to wield much bargaining power over others, as everyone would perceive the reputational situation differently. Relevant to this last point, Tomasello (2008) argued and presented evidence that early humans at this time did not have a conventional language but only the natural gestures of pointing and pantomiming (as noted above), with which it would be extremely difficult to communicate complex narrative information about partners' past behaviors. In addition to all of this, if early humans were as concerned about the fate of their partners as we are picturing them here, then an individual would not want to wield her bargaining power over good partners to their detriment. In general, then, all individuals in this more restricted marketplace, no matter their skills, would know that they had bargaining power generally similar to that of the various partners on whom they depended.

In the more information-poor, egalitarian marketplace we are envisioning here, then, the important thing was simply to find a partner with whom one could expect to be successful, and these were moderately plentiful with more or less equal bargaining power. Given individuals who could not help but view a collaborative partner as in some sense equivalent to themselves, the outcome psychologically would have been a respect for partners and, indeed, a *mutual respect* between partners (what Darwall [1977] calls "recognition respect"). This mutual respect would thus have had a strategic component—each partner recognized the equal bargaining power of the other—but it also would have had a nonstrategic component, based on the genuine sense of partner (self–other) equivalence that all individuals recognized as a result of their participation in and adaptation to joint intentional activities. Together, these two components—the strategic and the genuine—led to a "mutual respect between mutually accountable persons" (Darwall, 2006, p. 36). And so were born second-personal, cooperative agents who respected one another's equal status based on participation both in the collaborative activity itself (partner-independent role ideals led to a sense of self–other equivalence) and in the wider marketplace of partner choice (in which both had equal bargaining power).

Partner Control and Mutual Deservingness

In partner choice one attempts to locate and engage a good partner. In partner control one attempts to improve the behavior of a less than fully cooperative partner, for example, by punishment or other means. In a limited pool of potential partners, being able to turn a bad partner into a good one would have had obvious adaptive advantages.

An especially important situation for partner control in early human collaborative foraging would have been attempting to control free riders. And this is not a given. In chimpanzees' group hunting of monkeys, free riding is prevalent—individuals who contributed nothing to the group hunt still get meat—and participants do not attempt to control it. Boesch (1994) reports that chimpanzee individuals get more meat when they are actually in the hunt than if they are bystanders, but bystanders still get plenty. This distribution is most likely due not to any kind of reward for participation but simply to the spatial fact that hunters are nearer to the captor when the capture is made and so join into the consumption more quickly, whereas bystanders arrive later and then must beg and harass their way to meat from outside the immediate circle of consumers. This interpretation is supported by a recent experiment designed to mimic this exact situation. The finding was that chimpanzee "captors" of the spoils of a collaboration did not share more with other collaborative contributors than with noncontributors, if it was experimentally controlled how close or far individuals were to the "kill site" (Melis et al., 2011a). In contrast, human children as "captors" in a similar situation shared more with collaborative contributors than with noncontributors, regardless of how close or far away they were at the time of capture; they actively excluded free riders (Melis et al., 2013).

In the early humans we are picturing here free riding was not initially a problem, because the number of individuals available was the same as the number needed for foraging success—two—and so slacking off was self-defeating: if I do not do my job, then neither I nor my partner will get food. But others could come up after the kill, and they were essentially competitors—from outside the collaboration—so at some point humans also evolved the tendency to deter, and so to try to control, free riders by denying them a share of the spoils. One can imagine that such attempts at excluding free riders were the original acts of retribution: you did not participate in the collaboration, so you cannot share in the spoils. In this case all contributing partners are seen as equivalent or equal, but noncontributors are not.

The process of excluding free riders opens the door to the dividing of resources on the basis of some kind of deservingness. The simplest distributional scheme is that participants get more or less equal shares and nonparticipants get nothing, and indeed, young children have a very strong tendency to divide the spoils of a collaboration equally (Warneken et al., 2011; Hamann et al., 2011; Ulber et al., submitted). But in extreme cases, gradations of participation might also be taken into account. For example, an individual who arrived late and joined into the collaboration for just the final part—but in a crucially important way—might merit some of the spoils. And children will in some extreme cases distribute resources among collaborative partners on the basis of work effort and contribution (Hamann et al., 2014; Kanngiesser and Warneken, 2012). In the current evolutionary hypothesis, exclusion from the spoils (or part of the spoils) would have been the only immediate punishment meted out to laggards and free riders initially, and one could certainly call this exclusion process strategic. But, as we have stressed, early humans at this point also had a nonstrategic recognition of the equivalence or equality of partners, including most importantly self and other. Once again, then, we may picture a combining of strategic and nonstrategic considerations—the strategic exclusion of free riders and the genuine recognition of partner (self–other) equivalence—leading to something like an emerging sense of the relative *deservingness* of individuals in sharing the spoils based on participation, or lack thereof, in the foraging event. Individuals who did not contribute do not deserve any of the spoils, and individuals who did contribute deserve equal shares of the spoils (with some room for gradations).

Cooperative Identity

Our account so far has noted two related but distinct social arenas. On the one hand, there are the dyadic interactions that early human individuals had with collaborative partners: "I" relating to "you" as parts of a "we." On the other hand, there is the larger pool of potential partners in the social group as a whole in situations of partner choice. In this larger social pool, what an individual needed was to be perceived as a competent collaborative partner, that is, to have good foraging partnerships with multiple individuals (in the absence of a fully public reputation). Another way of saying this is that the individual needed a social identity as a competent collaborative partner—a competent second-personal agent—who showed equal respect for other deserving partners.

As early human individuals were creating a social identity, they were simultaneously creating a personal sense of identity. As a specific application of their understanding of role reversal and interchangeability, early humans understood both roles in the evaluative process and how they interrelated: they evaluated partners and knew that they were being similarly evaluated. The result was that an individual could imagine herself in the place of another as she evaluated him, or conversely, she could imagine herself in the place of another as he evaluated her. The result was that, just as each of the partners came to feel that free riders did not deserve any of the spoils, so the individual could judge in a given situation that she herself did not deserve any of the spoils. This role interchangeability in the evaluation of partners linked one's social identity in the group as a competent collaborative partner to one's sense of personal identity. Early humans' sense of personal identity, then, was based on an essentially impartial judgment that did not favor the self over others. I judge myself as I would judge others because I cannot help but do so, given that I see myself and others as basically equivalent and I naturally reverse roles in all my judgments involving not only others but also myself.

The process of creating a cooperative identity began for early humans already in early childhood. Each child began by viewing with awe the powerful and competent adults doing difficult and interesting things. The young child had none of the competencies required to participate, and he knew it. As he grew older, he gained in the competencies and knowledge involved in specific collaborative activities. Implicitly, by attempting to participate, he petitioned the adults for recognition by inclusion (Honneth, 1995). And he knew that this recognition could come only from those whom he was petitioning, because theirs was the only judgment that mattered. (If one wants to be accepted as a competent chess player, recognition by one's mother matters little.) The developing child was seeking respect and recognition as a competent collaborative partner—as a second-personal agent deserving of mutual respect—and when he was granted this, it would become an integral part of his cooperative and so personal identity. (We will reserve the term *moral identity* for our next evolutionary step in the context of a moral community—not just a loosely structured social group—that governed all aspects of individuals' lives.) Because the individual needed to collaborate or else starve, the way that others viewed him—and his resulting sense of personal cooperative identity—was a matter of life and death.

As individuals became second-personal agents, and needed to be chosen as partners, it was important to signal to others both one's cooperative identity and one's recognition of their cooperative identity. A major way of doing this was through cooperative communication via so-called second-personal address. Thus, one opened the channel of cooperative communication by addressing the recipient with respect and recognition, and this address simultaneously asked for the same respect and recognition in return. Given this second-personal address and its acknowledgment, both communicator and recipient normally trusted one another. If the recipient perceived the second-personal address but then rebuffed it, that would constitute a serious breach of their mutual assumptions of cooperation, respect, and trust. Second-personal address therefore both presumed and helped to constitute what it was for an early human to be a competent collaborative partner. It also enabled the making of the kind of local, temporary, two-person contract known as a joint commitment.

Joint Commitment

The kinds of collaborative activities so far described were inherently risky, because they were based only on strategic trust. I think I know your motives, and I rely on that knowledge, but I could easily be wrong: perhaps you have eaten more recently than I thought and so are unmotivated, or perhaps you judge that the hunt will be more difficult than I thought and so you demur, or perhaps something unexpected and better comes along during the hunt that tempts you away. What we need to take the risk, then, is for each of us to trust one another more deeply, in a more committed way. What we need is for each of us to feel that we truly *ought* to follow through on our collaboration, that we truly *owe* it to one another.

It was the deepest insight of that most insightful of social theorists, Jean-Jacques Rousseau (later modified by Kant and the German idealists), that the only possible source for a personal sense of "ought" of this genuine kind is my identification with and deference to a larger (even idealized) social body of which I myself am a part. I freely grant authority—legitimate authority—over "me" to the supraindividual entity that is "we," and indeed, I will defer to that "we" to the point that if you rebuke me for nonideal behavior, I will join you in this rebuke (either overtly or in a personal feeling of guilt), judging that it is indeed deserved. I choose to identify most deeply in the situation with the supraindividual "we" over "me" because to do otherwise would be to renounce my cooperative and thus my personal identity (see Korsgaard, 1996a).

The original supraindividual entity that early humans were able to create was the "we" established by a joint commitment between two collaborative partners. There is ongoing theoretical debate about whether joint commitments are something additional to a joint collaborative activity structured by joint goals and intentions (Bratman, 1992, 2014), or whether the notion of a joint goal makes sense only as part of an existing joint commitment (Gilbert, 2003, 2014). We believe that there can be joint intentional activities into which two individuals simply "fall" and then form a joint goal given their common interests and common-ground strategic trust that each will act rationally in her own best interest (as elaborated above). Collaborative activities as initiated by joint commitments, in contrast, represent the special case in which we explicitly and openly express our commitment to one another as mutually respectful second-personal agents and so form a bond of "normative trust." Initiating a collaborative activity by means of an overt and explicit act of cooperative communication brings everything out into the open, so that the resulting joint commitment is underwritten and backed by the signatories' cooperative identities. This joint self-regulation represents the kind of "we > me" control, as we may call it, that, when internalized, becomes a sense of second-personal responsibility to one's collaborative partner.

The Original Agreements

There are many different ways to begin a collaborative activity. Chimpanzees mostly rely on a leader–follower strategy. In hunting monkeys, for example, one individual typically begins the chase, and others join in. (This is also characteristic of most coalitionary contests, in which most often one individual is already fighting and its friends join in on its side.) In an experimentally constructed stag hunt game, virtually all pairs of chimpanzees adopted such a leader–follower strategy (Bullinger et al., 2011b). But this is risky for the leader because it is counting on the others to follow. Indeed, the most frequent leaders in group hunts are youngsters (Boesch, 1994), who might be either ignorant of the contingencies or else especially impulsive, and "impact hunters," who might be individuals that are especially impulsive, risk seeking, or overconfident.

The main way for individuals to lower their risk in a stag hunt situation is to communicate with their partner before abandoning their hare. In experimentally constructed situations with a high degree of risk, chimpanzees basically never do this, whereas human children do it quite often (typically with attention-getting gestures and informative verbal utterances announcing the arrival of the stag; Duguid et al., 2014). Therefore, given a stag hunt situation in which both partners are attempting to reduce risk, and given some initial skills of cooperative communication, one can imagine that early humans began using some kind of attention-getting gesture or vocalization to coordinate before forsaking their hare: either indicating one's own plan of action or suggesting one for the partner or the dyad. These kinds of communicative acts would have been extremely helpful to get things started, but obtaining a communicative response from the partner—perhaps even a commitment—would have been even better. Since each partner is in the very same risky situation, the best option for each is to obtain such a commitment from the other, and the result is a joint commitment.

For social theorists focused on normativity, joint commitments represent nothing less than the "social atoms" of uniquely human social interaction

(Gilbert, 2003, 2014). Joint commitments are both basic and essential because they explicitly acknowledge our mutual interdependence in the upcoming collaborative activity and seek to manage it. They assume that each party is a second-personal agent with a cooperative identity who can be trusted in the requisite way to live up to his role-specific ideal. To be trusted in the requisite way—what we will call normative trust—an individual not only must have certain cognitive and physical competencies but also must have a cooperative identity as a result of treating other second-personal agents with mutual respect, and judging individuals, including himself, for their deservingness of both rewards and punishments. And he must be willing to put this cooperative identity on the line.

To get started, joint commitments are created when one individual makes some kind of explicit communicative offer to another that "we" do X, and then that other individual accepts, either explicitly via her own cooperative communicative act or implicitly by just beginning to play her role (based on comprehension of the communicated offer). Initiating and accepting an invitation to a joint activity employs second-personal address, presupposing a mutual attitude of cooperation, and making the dyad's common-ground assumptions about their role-specific ideals wholly overt and in the open. Each individual intentionally invites the other, in the open, to make plans, even risky plans, around the fact that she will do X—to trust that she will persist in pursuing X, ignoring fatigue and outside temptations, until both of them are satisfied with the result (Friedrich and Southwood, 2011); to believe that she may proceed without fear of failure due to his lassitude or negligence (Scanlon, 1990); and, in general, to depend on him.[3] And crucially, joint commitments can be terminated only by some kind of joint agreement as well. One partner cannot just decide she is no longer committed; she must ask the other to end the commitment, and the partner must accept (Gilbert, 2011). Joint commitments are joint all the way down.

With linguistic creatures, the prototype is an offer begun by something like "Let's X" and accepted with "OK" (and ended with something like "Sorry, I have to X now. OK?" "OK."). But language is not essential; all that is needed is some kind of cooperative communicative act. Thus, when Warneken et al. (2006, 2007) engaged fourteen- and eighteen-month-old infants in a collaborative activity and then abruptly stopped interacting, the infants attempted to reengage the partner through communicative attempts, most often some kind of pointing or beckoning. That is to say, they addressed the partner second-personally, implicitly offering "Let's . . . ," and expecting, if not demanding, a response of some kind. As evidence for this interpretation, Gräfenhain et al. (2009, study 1) had an adult form a joint commitment with some three-year-old children—the adult suggested "Let's X," and the child explicitly accepted—but with others the collaborative interaction was begun by the adult simply following into the child's activity. Then the adult abruptly stopped interacting. Children who were party to the joint commitment were much more likely than the other children to try to reengage the recalcitrant partner. These children seemingly reasoned: if "we" have a joint commitment, then "you" ought to continue as long as needed.

And children know what joint commitments mean for their own behavior with a collaborative partner as well. Thus, in a recent experiment three-year-old children committed to a joint task, but then, unexpectedly, one child got access to his reward early. For the partner to benefit as well, this child had to continue to collaborate even though there was no further reward available to him. Nevertheless, most children eagerly assisted their unlucky partner so that both ended up with a reward—and more often than if the partner just asked for help in a similar situation but outside of any collaboration or commitment (Hamann et al., 2012). (In contrast, when pairs of chimpanzees were tested in this same situation, as soon as the first one got its reward, it abandoned the other and went off on its own to consume it [Greenberg et al., 2010].) In a follow-up study, Gräfenhain et al. (2013) found that pairs of three-year-olds who committed to work on a puzzle together did such things as wait for their partner when she was delayed, repair damage done by their partner, refrain from tattling on their partner, and perform their partner's role for her when she was unable (i.e., more than did pairs of children who simply played in parallel for the same amount of time). When young children make a joint commitment with a peer, they help and support her much more strongly than when they are just playing together.

Second-Personal Protest

The content of the joint commitment is thus that each partner plays her collaborative role diligently and in the ideal way until both have benefited. But what happens if one partner does not? The answer is that he gets sanctioned, and, of crucial importance, it comes from "us." That is to say, it is of the essence of joint commitments that "we" agree to sanction together whichever of us does not fulfill her role-specific ideal. This gives the sanctioning a legitimate, socially normative force that acts as a self-regulatory device to keep the joint activity on track despite individual temptations and other outside threats. The normative force of the joint commitment is thus both the positive force of equal respect that each partner feels for the other—my respected partner deserves my diligence—and the negative force of legitimate sanctions, deserved sanctions, for reneging. To reduce their risk, then, *each partner to a joint commitment gives to the other the authority to initiate sanctioning when, by their common-ground standards made explicit via the joint commitment, it is deserved.* The role reversal in evaluative judgments is key here, as a derelict partner also judges himself as deserving of resentment and sanctioning. Each party to a joint commitment thus grants to the other what Darwall (2013) calls the "representative authority" (representing our "we") to call the other to task with respect to any deviations from their common-ground role ideals.

As part of this process of joint self-regulation, early humans forged a creative synthesis of partner choice and partner control—which also employed cooperative communication—that we may call second-personal protest (also called by Darwall [2006], Smith [2013], and others such things as "legitimate protest" and "moral protest"). To illustrate, let us return to the experimental situation in which two individuals collaborate to pull in a board with one pile of food in the middle. Chimpanzees dealt with this situation based primarily on dominance: if the subordinate attempted to take the food, the dominant attacked it, and if the dominant attempted to take the food, the subordinate just let it (Melis et al., 2006b). In contrast, three-year-old children typically did neither of these things, although they were perfectly capable of doing either. Instead, what happened most often was that if a greedy child attempted to take all of the sweets, she was met with a protest (Warneken et al., 2011). The aggrieved child expressed resentment toward the greedy child's actions, for example, by squawking loudly at the greedy child or saying "Hey!" or "Katie!" Critically, if neither child took more than half of the sweets, there was no such protest, and when there was protest over an unequal grab, the greedy child almost always relented.

Unlike chimpanzees, then, young children most often respond to greediness after collaboration with an act of cooperative communication initiated via second-personal address—expressing resentment. But what exactly does the protesting child resent? From their subsequent actions it is clear that the protester was focused not on getting more food but, rather, on getting an equal amount of food—he protested only when the split was unequal—and this is how the greedy child understood it. The protest was thus premised on the children's common-ground understanding of the ideal behavior of collaborative partners in dividing the spoils—equality—which the greedy child just violated. And so it is not just about getting more food, in which case the child should be protesting any split other than 100 percent for herself; it is rather about the individual getting what she deserves. In the words of Adam Smith (1759/1982, pp. 95–96), the aim of resentful protest is "to make [one's partner] sensible, that the person whom he injured did not deserve to be treated in that manner." Second-personal protest is thus a cooperative and respectful response to the offender's disrespectful actions; it does not seek to punish the partner directly, only to inform her of the resentment, assuming her to be someone who knows better than to do this (i.e., to treat others as less than equals). The offended child thus makes a second-personal demand: you must acknowledge that I should not be treated as less than equal.

But why should the offender respect this claim? The reason is that it is addressed to her second-personally with an assumption of cooperation, and if she is indeed cooperative, and wishes to remain so, then she needs to respond appropriately (Darwall, 2006, 2013). Second-personal protest is thus a proleptic act: it treats the offender respectfully as a competent cooperative partner, and if she lives up to that by respecting the claim, then she keeps this identity; if not, then she is at risk of losing her cooperative identity. The attempt at partner control represented by second-personal protest is thus implicitly backed up—in the event that the offender compounds the disrespect by disregarding the protest—by the threat of exclusion via partner choice: if you do not shape up, I will ship out. But it seldom comes to that because, first of all, the offender cannot help but see the protest as in some sense legitimate or deserved: the partner whom she has deprived and disrespected is an individual just like her, equivalent to her, who has contributed to producing the spoils equally with her, and so she does not deserve a less-than-equal share of the spoils. And then second, on top of this, there is the joint commitment in which both partners openly place their cooperative identity on the line (foreswearing the excuse of ignorance): if I behave badly I will engage in a role reversal judgment on myself and affirm your censure as deserved.

It is telling that second-personal protest as a communicative act can be totally empty referentially; a simple "Hey!" or a squawk will do the job. Complex language is not necessary because it is in the partners' common-ground understanding that people should not be treated disrespectfully in this way, and so all that is needed is second-personal address expressing resentment. Thus, children in these experiments often say things like "I only have one," which assumes as common-ground knowledge that things should be otherwise, and in this context we both know what that otherwise should be. The cooperative premise of the second-personal address is that you will want to know why I am communicating, namely, in this case, to express my resentment, which you will want to do something about before it escalates. In not specifying what you should do but only registering a protest, I am treating you as a cooperative agent who only needs to be reminded of what you already know you should do.

Given its demand for equal treatment, second-personal protest may be seen as the most explicit implementation of the recognition of all cooperative partners as equally deserving individuals, that is, as second-personal agents with the requisite status for entering into joint commitments. It is performed by the partner who has been aggrieved—which presumably is most natural—but it assumes that the aggriever will, of her own free will, recognize the validity of the claim and rectify the situation. And indeed, this is the most general point that Darwall (2013) believes is missed in classical accounts of human morality (he focuses on Hume): they fail to recognize the interpersonal nature of the human sense of fairness and justice. In many classic accounts the individual approves or disapproves of the behavior of others (perhaps from a "general point of view") based ultimately on sympathy for those who might be harmed (or possibly with the group and its functioning). But the structure of second-personal protest shows that, at least in many cases, things are not driven by sympathy for the harmed but, rather, by resentment against disrespect, resentment against being treated as something less than equal. I protest directly to you—with resentful second-personal address demanding respect—and I expect you to legitimate my claim and react appropriately. The structure is thus not of a private judgment simpliciter—though it in some sense includes that—but, rather, of one side of a dialogue that constitutes our interpersonal interaction as mutually respectful partners.

Dividing the Spoils Fairly

And what about dividing the spoils of a collaborative activity structured by a joint commitment? As documented above, even very young toddlers freely share resources when they have collaborated to produce them, and this sharing almost always results in equality between partners (e.g., Warneken et al., 2011). But in those studies the children did not have to give up anything in their possession, only to refrain from hogging unowned resources. This might conceivably have been done simply out of fear of conflict with the partner.

Hamann et al. (2011) made things more difficult for children. In their study, pairs of three-year-olds always ended up in a situation in which one of them had three rewards (the lucky child) and the other had only one (the unlucky child), so that to create an equal distribution the lucky child would have to sacrifice. What differed across three experimental conditions was what led to the asymmetrical distribution. In one condition, the unequal distribution resulted from participants simply walking into the room and finding three versus one reward at each end of a platform. In this condition, the children were selfish: the lucky child almost never shared with the partner. In a second condition, each child pulled his own separate rope, and this resulted in the same asymmetrical rewards. In this condition, the lucky child shared sometimes. But in a final condition, the asymmetrical rewards resulted from an equal collaborative effort on the part of the two children pulling together. In this case, the lucky child shared with the unlucky child (to create an equal 2:2 split) almost all of the time! Presumably, they felt that if they both worked equally to produce the rewards, they both deserved them equally. (Notably, when the same experiment was run with chimpanzees, they hardly shared at all, even when all they had to do was not to block the other from accessing the extra reward, and not differently across conditions.) Three-year-old children will actually give up a resource (show an aversion to positive inequity) in order to balance things with, and only with, their collaborative partner.

The obvious interpretation is that, in the context of a joint intentional activity, young children feel that they and their partner both *deserve* an equal share of the spoils. (It could be argued that these children were only blindly following a sharing rule that they learned from their parents. But in this case, they should have divided the rewards equally in all three conditions—unless, implausibly, the rule was to share resources only after collaboration.) But why should collaboration have such a dramatic effect on young children's willingness to give up a resource they already possess in order to equalize with a partner? Presumably, once the children entered into the collaborative activity they became committed to the relevant behavioral ideals. Some children did actually make something like a joint commitment before they began pulling together, while others simply looked to one another and waited for eye contact before beginning, which, given their common-ground knowledge of how the apparatus worked, could be viewed as a kind of implicit joint commitment. In any case, something about collaboration seems to have engendered a "we" that led young children to see their partner as equally deserving of the spoils. It was this sense of equal deservingness, we would argue, that motivated three-year-olds to willingly hand over a resource already in their possession, which they would not otherwise do.

The claim that the children in these studies have a sense of deservingness—as opposed to, say, a simple preference—is crucial for any arguments about their morality. The essential point is that, although divvying up the spoils would seem to be about the resources involved, children's behavior in the experiments shows that it is actually about something else, something more interpersonal. The key observation is that people's satisfaction with a division of resources depends not on the absolute value of what they have received but on what they have received relative to other individuals (Mussweiler, 2003). Thus, in the experiments of Warneken et al. (2011), Hamann et al. (2011), and others, to create equality required at least one child to make a social comparison between what she got and what her partner got. The result that produced dissatisfaction—and indeed, resentment as expressed in second-personal protest in many cases—was not the absolute amount received but, rather, the amount received relative to the partner. The satisfaction of the partners was based fundamentally on a social comparison.

Following Honneth (1995), then, we may say that my resentment in such situations is not about the absolute amount of resources I receive—with which I would otherwise be satisfied—but, rather, about the disrespect you are showing me by taking more than me, that is, by considering me to deserve less than you. That is not fair. Indeed, I also do not think it is fair if I receive more than you because I genuinely see you as an equally deserving individual, and in addition, I made a joint commitment to respect my partner as an equal second-personal agent. Dividing the spoils fairly is thus not about an equality of "stuff"; it is about an equality of respect. This is why I am not just disappointed to receive less than you; rather, I am positively resentful. A joint commitment thus raises the stakes when it comes time to divide the spoils of our collaborative efforts. Partners to a joint commitment do not just prefer that we share equally; they feel that we owe it to one another to share equally.

Second-Personal Responsibility and Guilt

We argued above that role-specific ideals were the first socially constructed normative standards. And now, as a result of their joint commitments with one another, early human individuals felt responsible to their partners for living up to those shared standards. (NB: Because early humans felt responsible only to their partner in a joint commitment, we will say they felt a sense of second-personal responsibility to their partner; we will reserve the term *obligation* for the next step in our story involving a wider moral community.)

Strategically, an early human individual behaved cooperatively whenever he could in order to create and maintain a cooperative identity in the pool of potential collaborators, someone who could be trusted to keep his joint commitments. This gave a certain urgency to treating others cooperatively. But urgency is not "ought," or at least not a moral "ought." For that we need two new attitudes. The first is that I treat others cooperatively because that is what they deserve. I treat my partner as an equally deserving individual because . . . well . . . he is one. The second is that I identify with our joint agent "we," and it is actually this "we" that makes judgments of deservingness. These judgments thus do not represent what *"they"* think of me but, rather, what *"we"* think of me. I develop a sense of personal identity based on how "we," our joint agent, judges me. Since I, as representative of our "we," judge myself in the same way that I judge others, all of my judgments have an impartiality that gives them further legitimacy in the sense that they clearly do not represent my self-serving motives. Early humans' sense of second-personal responsibility—feeling that they "ought" to treat their partner cooperatively—was thus not strategic reputation management but, rather, a kind of socially normative self-regulation.

Guilt is essentially using this process of socially normative self-regulation to judge one's own completed actions. (NB: Because early humans felt guilty only within the confines of a joint commitment with a partner, we will say that they felt a kind of second-personal guilt.) It is not just punishing oneself, although it has an element of that; it is judging that "I ought not to have done that." The "ought" again means that the judgment is coming from something larger with which I identify—specifically, our "we"—and so I trust its legitimacy. Those who behave noncooperatively and then feel regret only out of a fear that others will discover it and punish them are not really feeling guilty at all. Feeling guilty means that my current self, as representative of the "we" that I formed with my partner via a joint commitment, believes that my noncooperative act deserves to be condemned. And an additional sense of legitimacy comes from the fact that I am judging myself impartially, in the same way that I judge others, for noncooperative acts: I condemn whoever acts in that way, including myself.

The normative (and not just strategic) force of guilt is apparent overtly in the way it leads individuals to attempt to repair the damage they have done. For example, in a recent experiment, when three-year-old children felt sympathy for another child whose toy had broken, they did not attempt to repair it very often (even though this might have been appreciated). But when they themselves had inadvertently broken the other child's toy, they went to extensive efforts to repair it (and much more than if they had broken the toy when it caused no harm to anyone) (Vaish et al., in press). Guilt, even of the second-personal variety, encourages a renunciation of the guilt-inducing act and attempts at reparation.

Early humans' sense of second-personal responsibility thus derived both from the positive motivation to live up to one's cooperative and personal identity—so that I may judge myself as someone who deserves respectful treatment as a second-personal agent—and from a kind of preemptive attempt to avoid the legitimate chastisement of my partner and myself. As one example, in a recent experiment Gräfenhain et al. (2009, study 2) had a child and an adult make a joint commitment to play a game together. Then, another adult enticed the child away to a new, more attractive game. In response, two-year-old children simply dropped everything and took off for the new game. But three-year-old children understood their joint commitment and acted responsibly; before taking off (if they did in fact take off) they hesitated and looked to the adult, and often did something overt to "take leave," for example, handing over the tool used in the game or even verbally apologizing (much more than in the exact same situation with no prior joint commitment). The children recognized that they had a joint commitment, and because breaking it would harm and disrespect their partner, they had a responsibility to her to acknowledge that they were breaking it and that they regretted it. Adults in similar situations always "take leave" because they know that to break a joint commitment they must ask permission and their partner must grant it. This is acting responsibly to one's partner before it comes to some kind of second-personal protest or sense of guilt for noncooperation.

Second-personal responsibility and second-personal guilt were thus the first socially normative attitudes of the human species, presumably deriving from a kind of internalization of the process of resentful second-personal protest. The individual, as representative of the "we" he had formed via a joint commitment, protested against himself for treating others in a way that they did not deserve to be treated. These attitudes were based partly on strategic concerns of maintaining one's cooperative identity. But at the same time they were also genuinely moral attitudes in the sense that the individual judged that he ought to behave responsibly with his partner because she deserved it, and also because the joint commitment specified that noncooperation from either partner ought to be condemned, impartially, because in fact it was deserved.

In any case, and at the risk of oversimplification, a schematic depiction of the most basic elements of a joint intentional activity structured by a joint commitment is presented in Figure 3.1.

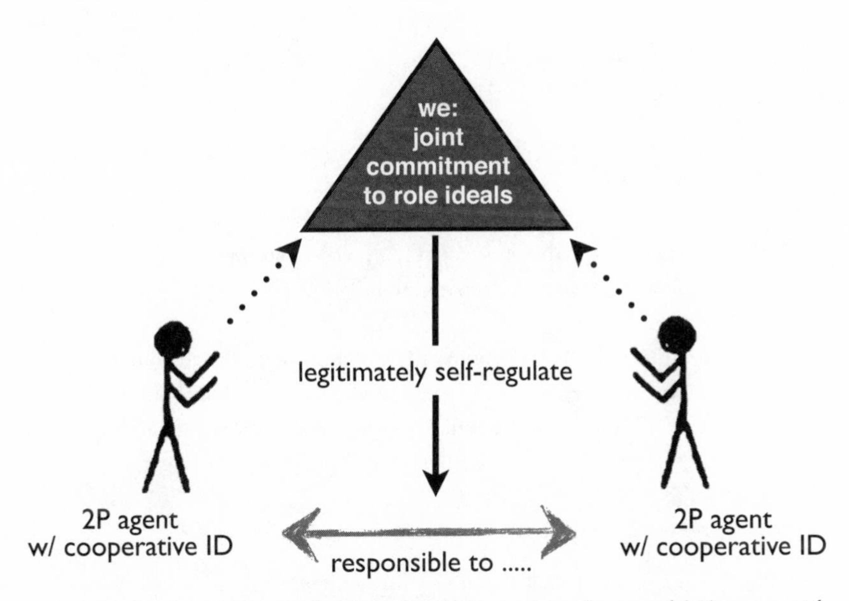

FIGURE 3.1. Joint intentional activity in which two second-personal (2P) agents with cooperative identities (ID) use their powers of joint intentionality and cooperative communication to make a joint commitment to create a supraindividual "we" based on normative trust (two dotted arrows pointing up to the triangle) that serves legitimately to self-regulate (solid arrow pointing down) the collaborative activity (horizontal, bidirectional arrow). The joint commitment is to act in accordance with their common-ground role ideals that define their and their partner's contributions toward joint success; that is, the commitment is to act responsibly toward their joint goal.

BOX 1. Human and Ape Collaboration

The claim is that early humans evolved some species-unique psychological competencies for collaborative foraging. A variety of experimental studies reported in this chapter comparing great apes and human children illustrate this more cooperative orientation and so provide at least indirect evidence for the first step in this hypothesized evolutionary story. The fact that the children in almost all of these studies were too young to be participating fully in the more group-minded normativity of their culture suggests a stage in the evolution of human morality that was mainly second-personal. Here is a summary of the most apposite of these studies. In all cases the results are summarized by statements of what young children do that other great apes do not do. In all cases, except when specifically indicated, these are explicitly comparative studies with methods designed to be as similar as possible for the different species.

Joint intentionality and cooperative communication: Young human children but not great apes

- form joint goals and joint attention with others, along with the individual roles and individual perspectives that these imply (Warneken et al., 2006; Tomasello and Carpenter, 2005);
- reverse roles in collaborative activities in ways that show a perspectival knowledge of the interchangeability of roles (Fletcher et al., 2012); and
- communicate cooperatively in order to coordinate the collaboration (Warneken et al., 2006; Melis et al., 2009), including to initiate collaboration in stag hunt–type scenarios (Duguid et al., 2014).

Dividing the spoils: Young human children but not great apes

- share the spoils of a collaborative effort even when these are readily monopolizable by one partner (Melis et al., 2006b; Warneken et al., 2011),
- share resources more equally with others if they result from a collaborative effort (Hamann et al., 2011), and
- prefer collaborative to solo foraging even when the payoffs are identical (Bullinger et al., 2011a; Rekers et al., 2011).

Partner choice and partner control: Young human children but not great apes

- share less of the spoils of collaboration with a free rider than with a partner (Melis et al., 2011a, 2013),
- help others in response to their immediately preceding help (Warneken and Tomasello, 2013; Melis et al., 2008), and
- modify their own cooperative and uncooperative behaviors depending on whether conspecific peers are watching (Engelmann et al., 2012).

Joint commitment: Young human children but not great apes

- are committed to a collaborative activity through to the end, even staying to help their partner after they have received their part of the spoils (Hamann et al., 2012; Gräfenhain et al., 2013; Greenberg et al., 2010);
- stay committed more when an explicit joint commitment has been made than when not, and take leave of others when breaking a joint commitment (Gräfenhain et al., 2009; there are no directly relevant ape studies); and
- protest respectfully when a partner breaks a joint commitment, and then the partner responds appropriately (Warneken et al., 2011; Melis et al., 2006b) and even feel guilt if they themselves harm another (Vaish et al., in press; there are no directly relevant ape studies).

These empirical facts establish beyond a reasonable doubt, in our opinion, that human beings are biologically adapted for collaboration in a way that other great apes are not. It is possible that some of these differences may be due to parents teaching or modeling for their children cooperative ways of doing things, and so the normal ontogenetic pathway incorporates parental input. But (1) great ape parents do not seem to be inclined to teach their children cooperation in this same way, and (2) our speculation is that, while parental teaching and modeling strongly influence the ontogeny of cooperation later in development, for these early emerging skills it is not necessary. (And the Hamann et al. [2011] study of sharing after collaboration, as noted above, does not lend itself to parent socialization explanations at all.) More extensive cross-cultural research involving cultures that vary in the prevalence and importance of the parental modeling and teaching of cooperation would help settle the issue.

The Original "Ought"

Classic social contract theorists such as Hobbes and Rousseau imagined isolated human individuals coming together and agreeing on a social contract for a full civil society. Neither they nor anyone else took this to be a description of actual historical events, and indeed, today contractualist moral philosophers are wont to say that the structure of societies is "as if" there were an agreement. But what we are positing here is an initial evolutionary step beyond great apes (who are social but not very cooperative and so, to some degree, isolated individuals) in which early humans created new supraindividual social structures by actual explicit agreements. These supraindividual social structures are a very long way from full civil societies—they are only local and temporary joint commitments between two collaborative partners—but they are nonetheless real and full agreements. Early humans collaborating in joint intentional activities, initiated and self-regulated by explicit joint commitments, thus represent the first evolutionary step in a natural history of the human social contract.

The goal in this chapter has been to describe and explain how this first step in the human social contract came into being. Most basically, we have attempted to characterize the emerging moral psychology of early humans from around four hundred thousand years ago as a new set of psychological adaptations for functioning within a new set of social conditions. The claim is not just that this was a modest first step on the way to a fully modern human morality but, rather, that this was the decisive moral step that bequeathed to modern human morality all of its most essential and distinctive elements.

Early Human Moral Psychology

We have characterized early human morality in terms of psychological adaptations for a particular way of life: obligate collaborative foraging with partner choice. The fundamental change from other great apes was that early human individuals became more dependent on others just as others became more dependent on them—they became more interdependent—in producing their life-sustaining resources. We have argued that the resulting moral psychology was a genuine morality, in the sense that individuals often had for their proximate goals helping others and treating them as they deserved to be treated, that is, fairly. These new moral attitudes were essential in helping to constitute a new form of rationality, cooperative rationality, by means of which early human individuals made sense of their newly cooperative world and made appropriate behavioral decisions about how best to navigate it.

We may schematize the genuinely moral dimensions of early human social interaction in terms of the following three formulas:

you > me

you = me

we > me

In all three formulas the self, "me," is either equated or subordinated to some other agent. These genuinely moral attitudes did not automatically win the day in any given decision-making event, of course, but their existence as powerful forces in early human psychology meant that, for the first time, individuals at least had the possibility of making a genuinely moral decision.

You > Me. The "you > me" formula represents the fundamental attitude involved in early humans' morality of sympathy. If the costs were not too great, early human individuals, just as other great apes, helped one another. But whereas other apes helped only kin and friends, early humans in addition began to help their (potential) collaborative partners, regardless of their past relationship; it was all business. Because this helping was based on the logic of interdependence, they did not help others with just any frivolous goal; rather, they targeted their welfare paternalistically (perhaps based on Smithian empathy) to help their potential partners remain in good shape for future collaborative activities. Because the helper herself benefited from these acts on the evolutionary level, some theorists would deem this sympathetic motive an illusion. But that is not true at the psychological level, where early humans' sympathetic concern for the welfare of others was pure.

You = Me. This formula represents the fundamental cognitive insight underlying early humans' morality of fairness. It derived from the dual-level cognitive structure of joint intentionality and the new social order it created. As an adaptation to problems of social coordination in collaborative foraging, the dual-level structure of jointness and individuality comprised the joint agent "we" as constituted by the collaborative partners "you" and "me" (perspectivally defined). Each partner had a role to play if the collaboration was to be successful, and over time particular dyads came to a common-ground understanding—from a "bird's eye view"—of the role ideals that needed to be fulfilled in particular collaborative activities for joint success. The bird's eye view on the collaborative process—entailing that roles could be interchanged among partners with no change in the ideal standards at work—led to a recognition of self–other equivalence in the collaborative process. This recognition was not itself moral, but it formed the basis for a kind of impartial stance—the self as only one in the partnership, subject to the same evaluative standards as the partner (see Nagel, 1970, 1986)—that was foundational in the emergence of a morality of fairness. In contrast to the case of sympathy based on interdependence, thinking of others as equivalent to oneself was not a motivation and was not directly selected on the evolutionary level at all; it was simply a part of the cognitive structure of joint intentionality adapted for effective social coordination. As such, from the point of view of morality, we may think of the recognition of self–other equivalence as a kind of structural "spandrel" that basically framed the way that individuals thought about things.

But in combination with other emerging motivations and attitudes, the recognition of self–other equivalence was crucial in constituting the most fundamental attitude in the morality of fairness: the sense of mutual respect and deservingness among (potential) partners. In the context of partner choice for a collaborative foraging activity, early human individuals understood their interdependence in the situation—and knew that their partner did as well—so that all partners had the bargaining power and standing to demand good treatment. Coupled with the cognitive insight of self–other equivalence, this bargaining power and standing led to a mutual respect between (potential) collaborative partners. Related to this, early humans needed some kind of control over free riders. Again coupled with the cognitive insight of self–other equivalence, individuals came to feel a sense of the equal deservingness of collaborative partners, as opposed to free riders, in sharing the spoils. As a result of both of these developments, early human individuals became second-personal agents with the standing to demand respect and their just deserts from their partner. Although there were no publicly shared reputations at this point, each individual did develop a cooperative identity (internalized into a personal identity) as someone whom specific other partners could trust. Failing in this regard elicited resentment from others and so damaged one's cooperative and personal identity.

We > Me. Finally, the "we > me" formula represents the way that second-personal agents made with one another a joint commitment to collaborate and thus freely relinquished overall control of their individual actions to the joint agent "we." Because the joint commitment was out in the open in cooperative communication, individuals could not deny that they made this commitment or that they knew how to fulfill it. The joint commitment created a supraindividual social structure, "we," that legitimately self-regulated the collaborative interaction, backed by the possibility of legitimate second-personal protest and sanctioning. The supraindividual entity created by "us" to regulate "us" thus reflected the Rousseauean capacity to bind oneself to a "we" both positively, aspiring to my virtuous cooperative identity, and negatively, avoiding legitimate sanctioning from "us." Internalizing this self-regulatory process constituted a sense of second-personal responsibility to one's partner, enforced by a genuine sense of second-personal guilt when one failed to live up to one's role.

It was thus rational, cooperatively rational, for early humans to make joint commitments with one another in the context of obligate collaborative foraging with partner choice. In this interdependent context, I care about my partner ("you"), and so it makes sense to surrender some control in the decision-making process ("we > me") to maximize the benefits all around. And in the shadow of the future, I must do everything I can to maintain my cooperative identity with each and every partner. But, in addition, and nonstrategically, partners to a joint commitment viewed one another with mutual respect as equally deserving second-personal agents (in contrast to free riders). This mutual respect was based at bottom on the individual's recognition of self–other equivalence, a recognition that had nothing to do with strategy, only with reality. This recognition was inert without some kind of motivational activation emanating from some particular social-interactional context, but nevertheless, the recognition of self–other equivalence structured the way that individuals understood what they were doing and should do. When two early humans made a joint commitment to collaborate, they genuinely believed that each of them should live up to their role ideals; their partner deserved that.

And whoever reneged, including themselves, legitimately deserved—from an impartial perspective—to be condemned.

The cooperative rationality we are positing here is the ultimate source of the human sense of "ought." Early human cooperative rationality expands human pro-attitudes to include the welfare of others, it presupposes second-personal agents who consider one another as equally deserving of respect and resources, and it focuses on the individual decision making that takes place within the context of the joint agent, "we," formed by a joint commitment. These new elements in the decision making of individuals created a socially normative sense of "ought" that was not just a preference or an emotion, but rather the dynamic force behind their actions. This is clear because individual decision making is characterized by a kind of pressure, an instrumental rational pressure, to act in a certain way: if my goal is to obtain honey, and I know that using this particular tool will accomplish that goal in the current circumstances, then it makes sense for me to use this tool (I ought to do it). The proposal is that in adapting to obligate collaborative foraging with partner choice early humans created a new set of social circumstances—a new social order, based on processes of joint intentionality and second-personal agency—in which it made sense to act morally. They thus developed a kind of second-personal responsibility to their collaborative partners—the original "ought"—that was not just a blind emotion or preference, but rather a sense of cooperative rational pressure that innervated their decision making.

Our three different forms of self-subordination—"you > me," "you = me," and "we > me"—thus made sense; they were rational, in the context of a way of life in which individuals were immediately and urgently interdependent with one another for their most basic needs. They were cooperatively rational given such specific cooperative challenges as choosing a good partner, making sure one is chosen as a partner oneself, recognizing behavioral ideals of role performance for both partners, responding to a partner's request for help, soliciting the trust of others in a joint commitment, sharing the spoils of the collaboration in mutually satisfactory ways, evaluating the actions of the partner and oneself, trying to control the actions of the partner, responding appropriately to partner protest, excluding free riders from the spoils, and correcting one's own behavior preemptively in acts of impression management and/or second-personal responsibility. Deciding how to act in these and similar situations could be done either strategically, considering only one's own interests, or morally by taking into account the interests of a partner of equal deservingness and the joint agent one formed with him. And so were born genuinely, if only locally, moral beings.

But Can There Be a Purely Second-Personal Morality?

What we have been attempting to characterize here is a purely second-personal morality. This is not the same goal as that of other contemporary theorists of the second-personal. For example, Darwall (2006, 2013) and others attempting to account for contemporary human morality in second-personal terms naturally assume individuals living in a cultural group governed by all kinds of cultural (and perhaps religious and legal) norms. Darwall thus says that when an individual makes a moral claim on another—for example, in moral protest—she is doing so with respect to mutually agreed-upon social norms emanating from and applying to the moral community at large. Darwall does not directly address the possibility of a second-personal morality outside of a cultural context.

Strawson (1962, pp. 15–16) thinks that the possibility of a purely second-personal morality is barely imaginable because he thinks this means a world of "moral solipsists," each of whom thinks that she is the only author and target of reactive attitudes, which is almost a contradiction in terms. But this is where joint intentionality plays its crucial role. The early humans we are picturing here cannot be solipsists of any kind because they are constantly entering with others into various kinds of "we" relations, including joint commitments. And within these "we" relations they are not just acting and reacting to one another as individuals with personal preferences and attitudes; rather, they are holding one another accountable with respect to more or less impartial standards independent of either of them. The impartial standards to which individuals bind themselves via a joint commitment represent the crucial third element—the external arbiter—that many social theorists believe is necessary for the notions of fairness and justice to exist in the first place (e.g., Kojève, 1982/2000).

We would thus argue that if the essence of a moral relationship is individuals committing to one another with respect to some mutually known and impartial normative standards, the early humans we are picturing here did indeed enter into moral relationships. It is just that the normative standards were only the role-specific ideals inherent in the specific joint intentional activities of the dyad, and the commitment was only local and temporary. The fact that two- and three-year-old children—who are not yet interacting socially in meaningful ways beyond the dyad—are operating in this way by protesting unequal treatment, and respecting such protest from others, offers an existence proof that there can be something like a purely second-personal morality without reference to more general cultural norms. Early humans were thus creatures who interacted with one another directly and second-personally in moral ways; what they did not do was intervene in the social interactions of third parties that did not concern them more or less directly. Their morality was thus limited and local, but, we would argue, it was a genuine morality nonetheless. It was clearly not a full-blown, group-minded, cultural morality of "objective" right and wrong applying to everyone in all situations. Rather, it was a morality for just one particular type of social activity—albeit one of immediate, urgent, and continuous importance—and this meant that in many other daily activities early humans would still have been very much ape-like. But soon, with the advent of modern human cultural life, this more limited and local way of operating would be totally transformed.

4

"Objective" Morality

> An obligation always takes the form of a reaction against the threat of a loss of identity.
>
> —CHRISTINE KORSGAARD, *THE SOURCES OF NORMATIVITY*

Not long after early humans transitioned into modern humans, somewhere in Africa, sometime around 150,000 years ago, they began forming separate and distinct cultural groups that competed with one another for resources. Interdependence now reigned not just at the level of the collaborating dyad, and not just in the domain of foraging, but at the level of the entire cultural group, and in all domains of life. The cultural group thus became one big collaborative enterprise that all of its members needed to do well so that they could do well. Outsiders to the group were essentially free riders or competitors—barbarians—and so excluded from the cultural collaboration.

The challenge for modern human individuals was to scale up from a life based on interdependent collaboration with well-known partners to a life lived in a cultural group with all kinds of interdependent groupmates. Cognitively, what was needed were skills and motivations not just of joint intentionality but of collective intentionality. These skills, along with newly powerful skills of cultural transmission, enabled individuals to create among themselves various types of conventional cultural practices, shared in the cultural common ground of the group. The roles in conventional cultural practices were fully agent independent: their role ideals were what anyone who would be one of us (i.e., any rational person) would need to do to promote collective success. At some point, these maximally generalized ideal standards came to be conceptualized as the "objectively" right (not wrong) ways to perform the role, including the generic role of simply being a contributing member of the culture. To participate in a conventional cultural practice—including when it became formalized into an institution—what the individual had to do most urgently was to conform to this right way of doing things. Some conventional cultural practices and their associated roles concerned things about which individuals already had second-personal moral attitudes; that is to say, they concerned potential issues of sympathy and fairness. In these cases, the normative role ideals specified not just conventional right and wrong but moral right and wrong.

This part of the scaling up from collaboration to culture was relatively straightforward: everything went from dyadic and local to universal and "objective." What was not so straightforward was the scaling up of joint commitments. The issue was that, unlike the socially self-regulating structure created by a joint commitment, for modern humans the largest and most important collective commitments of their culture—its conventional practices, norms, and institutions—were things that individuals did not create for themselves: they were born into them. The individual therefore faced, in theory, the problem of the social contract and its legitimacy. In practice, however, individuals naturally saw the self-regulating collective commitments into which they were born as legitimate because they identified with their cultural group; they assumed a kind of coauthorship such that the commitments were made by "us" for "us." In the case of moral norms, this legitimacy was fortified by its connection to second-personal morality. As they made moral decisions over time, modern human individuals created a kind of moral identity that was fostered and maintained by conforming to moral norms, by feeling guilty when they did not conform, and by justifying any nonconforming act to others and the self by grounding it in the group's shared values. The cultural rationality of modern humans was thus to freely relinquish much control of their individual actions to an unreflective conformity to the conventions, norms, and institutions of their group, with autonomous decision making (short of opting out completely) mostly confined to resolving conflicts between norms.

This chapter, then, attempts to explain how early humans' natural second-personal morality, from around 400,000 years ago, became modern humans' group-minded "objective" morality, starting around 150,000 years ago. Analogous to our account of early humans' second-personal morality, we characterize modern humans' "objective" morality in terms of something like a morality of sympathy: how the individual's dependence on the cultural group led to a special concern for and a loyalty to the group. Then, with respect to a morality of justice, we once again characterize the three sets of psychological processes involved. First are new cognitive processes of collective intentionality creating "objective" normative ideals of right and wrong. Second are new social-interactive processes of cultural agency with respect to the group's conventions, norms, and institutions. And third are new self-regulatory processes of moral self-governance based on the sense of collective commitment and obligation that a person with a moral identity has to her moral community. We will have some things to say near the end of the chapter about how processes of cultural group selection played an important role in all of this as well.

Two important preliminaries: First, we will use the expression *modern humans* for the species *Homo sapiens sapiens* that began to emerge some 150,000 years ago. But we are focusing mainly on the 140,000-year period before agriculture and modern civil societies, when humans were still living exclusively as hunter-gatherers in relatively small-scale tribal societies. We will refer to modern humans in the last 10,000 years since agriculture, especially those living in contemporary civil societies, as *contemporary humans,* and we will have a few words to say specifically about them in a coda at the end of the chapter.

Second, in many cases we will say that the members of a particular culture insist that this is the way that "we" do things. We will also say that members of a culture insist that this is the way things are done, the right way to do them, the way that any rational or moral person would do them. These different ways of talking are all intended to be more or less equivalent because, again, the focus is on modern humans before the rise of civil societies. For these early modern humans, "we" in this group are humans, and the other similar-looking creatures that we sometimes see in the distance—or interact with cautiously and with little comprehension—are barbarians and so not really humans at all. "We" know the right way to do things; "they" do not. It is in this sense, from this internal perspective, that a cultural, group-minded way of thinking is "objective."

Culture and Loyalty

Obligate collaborative foraging based in early humans' loosely structured and relatively small social groups was an evolutionarily stable strategy—until it wasn't. The basic problem was that it was too successful: the population sizes of some groups grew until they started bumping into one another regularly, which led to group competition for resources. The modern human cultural group consequently became, in effect, a single, self-sustaining collaborative enterprise, a collaborative foraging party writ large, aimed at the collective goal of group survival, with each individual playing his or her division-of-labor role, including the role of being a competent and loyal group member in general.

In a hostile environment with competitor groups always lurking, and with subsistence activities requiring significant specialized knowledge and tools, the individual was basically totally dependent on the group. Given this dependence, the two most immediate and urgent challenges for individuals were (1) to recognize, and to be recognized by, all of their many in-group compatriots, even those they hardly knew; and (2) to help and protect, and to be helped and protected by, all of the in-group compatriots with whom they were interdependent, which meant, especially as division of labor increased, basically everyone in the group.

Similarity and Group Identity

As modern human groups began growing and expanding, they began to splinter. But at the same time competition among groups posed a serious threat to small groups; there was safety in numbers. The result was the so-called tribal organization of modern human social groups, emerging perhaps 150,000–100,000 years ago (see Foley and Gamble, 2009; Hill et al., 2009). In general, tribal organization emerges as groups grow and expand until they fission into smaller "bands," which nevertheless coalesce together into a single cultural group for some purposes, especially intergroup conflicts.

For the individual, a serious problem in this new social organization was "Dunbar's number." Drawing on many different lines of research, Dunbar (1998) argued and presented evidence that human individuals can have intimate knowledge of no more than 150 or so individuals at one time; any more exceeds their social memory capacities. This meant that modern human individuals personally knew everyone in their band but not in their culture, and so they needed some new way to recognize their in-group cultural compatriots, especially since they needed them in intergroup conflicts (and, in addition, approaching an outgroup barbarian by mistake might have been lethal). The interdependence and solidarity among early human collaborative partners who had personal experience with one another were therefore no longer sufficient; a new way of forming interdependent and trusting bonds with all in-group compatriots was needed. That new way was similarity. Contemporary humans have many and diverse ways of displaying in-group similarity, but it is likely that the original ways were mainly behavioral: people who talk like me, prepare food like me, and net fish in the conventional way—that is, those who share my cultural practices—are most likely members of my cultural group. At some point modern humans even began to actively display their similarity to others in the group, via such things as special clothing or body marking, to highlight their distinctness from those in neighboring groups.

In this context, conformity becomes a necessity. And indeed, compared with other primates, human children are clearly more motivated to conform. That is to say, they do not just socially learn instrumentally useful actions, as do all apes; they also copy others' actions precisely in order to be like them. This has sometimes been called "social imitation" to emphasize that the learner's motivation is to display his similarity to others and so to affiliate with them (Carpenter, 2006). Experimental research has thus established that

- human children are much more concerned than are other great apes to copy the exact actions of others, including arbitrary gestures, conventions, and rituals (see Tennie et al., 2009, for a review);
- human children, but not great apes, copy even irrelevant parts of an action sequence in so-called overimitation (Horner and Whiten, 2005); and
- human children, but not great apes, conform to others even in situations when they have to override a previously successful strategy to do so, so-called strong conformity (Haun and Tomasello, 2011, 2014).[1]

Such social imitation and strong conformity are not primarily social learning strategies for increasing personal success in problem-solving situations; rather, they are primarily social learning strategies for aligning oneself with others so as to show one's affiliation and perhaps group identity with them (Over and Carpenter, 2013). A particularly important finding in this regard is that human infants will selectively imitate individuals who speak their language over individuals who speak differently (Buttelmann et al., 2013).

Eventually, modern human individuals began to see their cultural membership as their cultural identity. In the contemporary world, for example, we see individuals sacrificing their lives for their country or ethnic group in wars, even waving a flag for the group as they go down, and they do this even when all that is at stake is keeping their ethnic group's identity alive and separate from that of other group(s). More trivially, but still tellingly, contemporary individuals do such things as paint themselves colors and chant for their sports team. And perhaps most telling of all, contemporary individuals feel collective guilt or pride when someone in their group does something especially heinous or praiseworthy, even when they themselves have done nothing. In a recent experimental study, even young children who were arbitrarily assigned to a group by being dressed in similar clothing and given a common group label (e.g., the "green group") felt a need to apologize and make amends for violations by an in-group (but not for an outgroup) member (Over et al., submitted).

And so with modern humans was born a second way to form a "we." Modern human individuals felt solidarity not only with interdependent collaborative partners, as already in early humans, but also with in-group members who resembled them in behavior and appearance. Interestingly, these two types of social solidarity were already recognized by Durkheim (1893/1984) in his organic solidarity (based on collaborative interdependence) and mechanical solidarity (based on similarity), and they also find their way into the two basic principles of group formation as established in modern social psychological research: interpersonal interdependence (based on joint projects) and shared identity (based on similarity and group membership) (e.g., Lickel et al., 2007).

In-group Favoritism and Loyalty

Modern human individuals identified with their cultural group because everyone in the group needed everyone else—they were interdependent—for all kinds of life-sustaining help and support, including protection from the barbarians across the river. "We Waziris are the people who do things like this and not like that, who dress like this and not like that, and who always stick together in the face of danger." Modern human cultures also showed the beginnings of a growing division of labor—for example, some individuals making all the spears used in hunting, and others specializing in cooking—that intensified interdependence even more, as each individual's survival depended on many others doing their jobs reliably. The interdependence of two individuals in a dyadic foraging party now became the interdependence of everyone in the culture as a whole, including even in-group strangers. As a result, it became important for individuals to show loyalty to the group and so to prove that they could be trusted, especially when it came to intergroup conflicts (Bowles and Gintis, 2012).

Modern humans' group-minded interdependence thus served to spread human sympathy and helping to all in the group, best characterized as a sense of loyalty to the group. As a consequence, there emerged in modern humans a distinctive in-group/out-group psychology. In-group favoritism accompanied by outgroup prejudice is one of the best-documented phenomena in all of contemporary social psychology (e.g., Fiske, 2010), and it emerges in young children during the late preschool and especially during the school-age period (Dunham et al., 2008). This in-group bias is evident in many different domains of activity, but most important for current purposes is morality. Much recent research has demonstrated young children's special prosocial behavior toward those who imitate or act in synchrony with them (i.e., similarity in behavior) or who are in their same "minimal group" (i.e., similarity in appearance, minimally established experimentally via similar clothing and/or a common group label):

- When someone imitates a young child, the child is then especially prone to help the imitator (Carpenter et al., 2013) and to trust her (Over et al., 2013).
- When someone speaks the same language as the child (and with the same accent), he prefers that person (Kinzler et al., 2009) and is more likely to trust her (Kinzler et al., 2011).
- When several preschool children act synchronously with one another, for example, dancing together to the same music, they subsequently help and cooperate with one another more than with other children (Kirschner and Tomasello, 2010).
- School-age children show more sympathy and helping toward individuals labeled as in-group than those labeled as outgroup (Rhodes and Chalik, 2013).

- School-age children expect and favor loyalty to the group from their in-group compatriots, whereas they expect and favor disloyalty to the group in outgroup individuals (Killen et al., 2013; Misch et al., 2014).
- Preschool children care more about how they are evaluated by members of their minimally established in-group than by outgroup members (Engelmann et al., 2013).

This tendency of modern humans to selectively help, cooperate, and trust those who behave like them, look like them, or are labeled with a common group name is so strong that it has led some theorists to posit homophily—the tendency to affiliate, favor, and bond with similar others—as the basis of human culture (Haun and Over, 2014). Although chimpanzees and other non-human primates live in spatially segregated groups and are hostile to strangers (Wrangham and Peterson, 1996), this hostility is not, as far as we know, directed at other groups qua groups, based on their group's distinctive appearance or behavioral practices.

Early humans' morality of sympathy for collaborative partners thus scaled up relatively straightforwardly into modern humans' loyalty to everyone in the cultural group with which they identified. The issue now, then, is the scaling up of early humans' morality of fairness to their collaborative partners into modern humans' morality of justice to everyone in their cultural group (or moral community). Our account again focuses on the three key types of psychological processes:

- cognitive processes of collective intentionality;
- social-interactive processes of cultural agency and identity; and
- self-regulatory processes of moral-self-governance and moral identity.

We deal with each of these, in turn, in the three sections that follow.

Collective Intentionality

Cognitively, what modern humans did to adapt to their new social reality was to transform a joint intentionality geared for dyadic collaboration into a collective intentionality geared for cultural collaboration. Thus, the dual-level structure now came to comprise, on one level, again, the individual, but now, on the other level, the group-mindedness and collectivity of all who shared a common cultural life. There was thus also a transition from seeing an equivalence between oneself and one's collaborative partner, as did early humans, to seeing an equivalence among all who would be a member of the cultural group, that is to say, all rational beings. The result was that a cultural group's conventional cultural practices—made possible by skills of collective intentionality—contained role ideals that everyone knew (and knew that everyone knew in cultural common ground) applied to anyone and everyone, which amounted to a specification of the "objectively" correct and incorrect ways to do things.

Conventionalization and Cultural Common Ground

The collaborative interactions of early humans required a good bit of personal common ground between the specific partners involved. When modern humans collaborated with others, in contrast, it was sometimes perforce with individuals with whom they had little or no personal common ground. In addition, they sometimes created collective group goals for larger-scale cultural activities, including group defense, which also involved individuals who were largely unfamiliar with one another. The problem was that coordinating collaborative activities with no personal common ground among participants was difficult. The solution was collective or cultural common ground, which led, in combination with a tendency to conform, to conventional cultural practices.

Collective or cultural common ground is the basis of—it is almost the definition of—culture. That is to say, the group life of early humans turned into the cultural life of modern humans precisely when individuals became group-minded both in terms of groupish motivations such as loyalty and trust for their compatriots, as we have just seen, and in terms of an epistemic dimension of shared skills, knowledge, and beliefs—cultural common ground. This cultural common ground meant that everyone in a group knew that everyone in that group had had certain kinds of experiences—and thus skills, knowledge, and beliefs—based on everything from a common physical environment to a common set of child-rearing practices through which each of them had passed. Knowing this meant knowing many important things about the minds and likely behavior of others, often without ever interacting with them directly. Indeed, Chwe (2003) argues that the main function of culturally sanctioned public events such as weddings, inaugurations, funerals, and war dances is to establish explicitly that something is common knowledge to everyone in the group.

Being able to correctly identify what skills and knowledge are in the cultural common ground of the group facilitates coordination with in-group strangers. Thus, if I tell an in-group stranger that next time it rains we meet at "the big tree" to hunt antelope, I can assume confidently that he knows which tree I mean. And I do not need to tell him which weapons to bring, as that is part of our cultural common ground as well. Over time, the result is what we may call conventional cultural practices that everyone knows that everyone knows how to perform. It is not just that everyone does something a certain way that is crucial for conventional cultural practices, but that everyone expects everyone else to do it that way as well, and expects the others to expect them to do it that way also, and so on (see Lewis, 1969). Theoretical attempts to characterize conventional practices solely in terms of conformity to precedent (e.g., Millikan, 2005), as well as empirical attempts to demonstrate conventions in nonhuman primates (e.g., Bonnie et al., 2007), miss this crucial sharedness aspect, on which the ability to coordinate with others is based (Tomasello, 2006; see also Grüneisen et al., 2015). Without a knowledge of sharedness, flexible coordination is not possible (see Tomasello, 2014, for an extended discussion).

Even very young contemporary children are able to make strong inferences about the knowledge of in-group strangers based on assumed cultural common ground. For example, Liebal et al. (2013) had three-year-old children encounter a novel adult clearly from their in-group. This in-group stranger then asked them sincerely "Who is that?" while the two of them looked together at a Santa Claus toy and a toy that the child had just constructed before the adult arrived. Children answered by naming the newly created toy, as they were confident that no one in the culture, not even someone they had never before met, needs to ask who is Santa Claus. In a contrasting experimental condition, if the in-group stranger seemed to recognize one of the two toys, children assumed it was Santa Claus. They behaved in this same contrastive manner for a number of other familiar everyday cultural objects, such as a toy car, in contrast to a just-made novel toy. Young children can thus infer, without direct personal contact, some of the things that a person must know if they are members of the cultural group. Another example is linguistic conventions: young children expect that in-group strangers will know the conventional name of an object, but not an episodic fact about that object, such as who gave it to me (Diesendruck et al., 2010).

Conventional cultural practices are based on cultural common ground in that they assume that anyone—perhaps within some demographic and/or contextual specification (e.g., all women, all net-fishers)—can play a particular role in a practice if she is privy to the appropriate cultural common ground. If one wants to net-fish successfully, one has to do it in the Waziri way, and everyone knows that means that the chaser does X and the netter does Y; it matters not who you are. Roles in conventional cultural practices are thus fully agent independent in that they apply to *anyone* who would play that role, where "anyone" is not just a large group of individuals but a designation in principle of anyone who would be one of us. This in-principle designation means that knowing certain things is in fact a constitutive part of one's cultural identity. This was also true, of course, of the linguistic conventions that modern humans began using to communicate within (and only within) their groups. Linguistic conventions are a part of the cultural common ground of the group (Clark, 1996), so anyone who wants to communicate with any group member must do so in the conventionally expected way, both as communicator and as recipient, and he can expect his interlocutor to be competent (and to expect her to be competent) in both of these ways as well.

An important consequence of modern humans' participation in cultural common ground and the agent-independent roles of conventional cultural practices was thus the ability to take a fully agent-independent perspective on things: the perspective of anyone who would be one of us, an "objective" perspective, which represents maximal generality beyond early humans' more limited partner-independent perspective from within the dyadic interaction. (Tomasello [2014] contrasts early humans' "view from here and there" with modern humans' "view from nowhere.") The "objective" point of view of collective intentionality thus structured new ways of relating to others adapted for life in a large-scale society in which personal relationships were not sufficient to keep everyone coordinated and satisfied in the many different venues in which they interacted. It did this by objectifying cultural roles, including the most general role of simply being a contributing group member, and by putting the self in its group-minded place as only one among many (Nagel, 1970, 1986).

Modern humans thus experienced as a social reality the dual-level structure of, on the top level, the cultural group as a whole (i.e., constituting any and all rational persons) and, on the bottom level, its individuals, including the self, who served as interchangeable (agent-independent) cogs in the conventional cultural practices that kept the culture going. An individual could on occasion forget all of this and stick with his own personal point of view, of course, but sooner or later the awareness of his interdependence with others and the group, along with his cognitive skills of collective intentionality, would force on him the impartial perspective that could take into account equally and impersonally the perspectives of everyone concerned. And so now we have the new cast of characters created by culture and collective intentionality: "I," "you," and "we"—conceptualized as "any rational person"—the *ménage à beaucoup* from which the modern human morality of justice has sprung.

Conventionally Right and Wrong Ways to Do Things

Where early humans had role-specific ideals in personal common ground for how each partner in the dyad should play her role for joint success, modern humans had in cultural common ground the correct and incorrect ways for performing the roles in conventional cultural practices. In a sense, this is just a quantitative difference: instead of role reversal and interchangeability between two partners, we have full agent independence within the practices of the cultural group. But it is also more than that. Conventional cultural practices as the correct way (not incorrect way) to do things go beyond early humans' ad hoc ideals that two partners created for themselves and that they could just as easily dissolve. The correct and incorrect ways to do things emanate from something much more objective and authoritative than us, and so individuals cannot really change them. The collective intentionality point of view thus transformed early humans' highly local sense of role-specific ideals into modern humans' "objective" standards of the right (correct) and wrong (incorrect) way to perform conventional roles. Such an agent-independent or "objective" point of view is not sufficient for judgments of fairness or justice, but it is necessary, as has been explicitly recognized in one way or another by moral philosophers from Hume with his "general point of view," to Adam Smith with his "impartial spectator," to Mead with his "generalized other," to Rawls with his "veil of ignorance," and to Nagel with his "view from nowhere."

The objectification process comes out especially clearly in intentional pedagogy (Csibra and Gergely, 2009). Intentional pedagogy is unique to humans

(Thornton and Raihani, 2008), and it very likely emerged with the advent of modern humans, since this is when clear cultural differences between neighboring groups first appeared (Klein, 2009). The prototypical structure of intentional pedagogy is an adult insisting that a child learn important cultural information (Kruger and Tomasello, 1996). Intentional pedagogy is thus explicitly normative: the child is expected to listen and learn. But just as important, it is also generic in the sense that its normative ideals are kind relevant. It is not just that we found this nut under this tree but that nuts like these are found under trees like these. It is not just that to throw this spear I or you must hold it with three fingers and a thumb but that to throw a spear like this anyone must hold it like this. The voice of intentional pedagogy is consequently both generic and authoritative: it is stating as an objective fact how things are or how one must do things, and its source is not the personal opinion of the teacher but, rather, some objective world of how things are. The teacher is representing the culture's take on the objective world: "it is so" or "one must do it so." Indeed, Köymen et al. (2014, in press) found not only that preschoolers use such generic normative language when they teach things to peers but that they often objectify their instructions to the maximum by saying not only that "one must put it here" but also "it goes here," and not only "one shouldn't do it like that" but also "that's the wrong way to do it."

This objective perspective on the correct and incorrect ways to do things is further fortified by a historical dimension. Cultural practices are not just how we Waziris do things now but how our people have always done them. Net-fishing in this way is not just what we in our group do, and so what you should do, but what our venerated ancestors have done forever. It ensures our survival as a people, and it distinguishes us from the barbarians across the river. It is the correct way to do things. As modern human individuals grew into their cultures, they were thus learning the "objectively" right ways to do things. And when they were explicitly taught these practices by adults, they did not experience this teaching as from individuals stating their opinion or giving their advice; they experienced it as the authoritative voice of the culture on how things are and how one must behave.

Cultural Agency

Collective intentionality and cultural common ground, as cognitive skills, thus facilitated coordination among the members of a cultural group by creating stable sets of shared expectations. But modern humans' large-scale societies also presented a variety of challenges to cooperation of a more motivational nature (Richerson and Boyd, 2005). The most basic challenge was that larger groups meant more possibilities for cheating since there were more interactions among in-group strangers who knew little of one another's histories. Relatedly, larger endeavors created problems of collective action due to increased opportunities for individuals to free ride on the efforts of others undetected (Olson, 1965). These new contingencies required cultural agents to have at their disposal much stronger mechanisms of partner choice and control for cooperativizing the more competitive aspects of cultural life. Enter social norms.[2]

Social Norms

Early humans controlled their noncooperative partners by protesting directly to them. But modern humans had many and varied roles in the many different activities of a complex society, and second-personal protest confined to direct communication would have been much too local and small-scale to be effective in these wider-scope activities with strangers. What gradually evolved was a set of expectations that everyone in the group shared in cultural common ground about how individuals must behave in various situations to be cooperative. In effect, such social norms conventionalized and collectivized the process of evaluating and protesting against noncooperators and thus scaled up second-personal partner control into group-level social control, with the aim of ensuring the smooth functioning of social interactions in the group. And so now modern humans knew in their cultural common ground not just the "objectively" correct and incorrect ways to perform the roles of conventional cultural practices but the "objectively" right and the wrong ways to be cooperative, to be moral, with one's compatriots.

Consistent with the idea of social norms as serving to cooperativize competition, the domains most universally covered by social norms across societies are those involving the most pressing threats to the group's cohesiveness and well-being, that is, those that bring out most strongly individuals' selfish motives and tendencies to fight: food and sex (Hill, 2009). Thus, how a large and bountiful carcass is to be divvied up among all the members of a cultural group is strictly governed by social norms that everyone knows in cultural common ground ahead of time, and this effectively preempts most squabbling (Alvard, 2012). Similarly, whom one can and cannot have sex with—for example, relatives or children or someone else's spouse—is also strictly governed by social norms in the cultural common ground of the group, again preempting potentially heated conflicts that could undermine the group's effective functioning. In effect, social norms anticipate competition in potentially disruptive situations and make it clear how individuals must behave in such situations in order to cooperate.

Ultimately, social norms are conventionalizations, and purely as such their aim is conformity. To illustrate, let us imagine a modern human celebratory feast. It is a conventional cultural practice specified in cultural common ground as various conventional ways of doing things. Now imagine that an individual wears unconventional clothes to the feast. Here he has not harmed anyone, but it is in our cultural common ground what one typically wears to a feast, and he has intentionally not conformed. Because we all assume that everyone identifies with their cultural compatriots and wants to affiliate with them, the inference is that the nonconformist does not want to identify or affiliate with us and, since he knows what we are thinking, he does not respect our evaluations. We now mistrust him because he does not favor or respect our group.

But let us now imagine that an individual steals off with most of the celebratory food. By this act he has harmed and disrespected many people, so this is clearly a violation of second-personal morality deserving of censure. But at the same time it is a lack of conformity to the group's social norms: we Waziris do not steal food from a celebration, and anyone who does so either is not one of us or does not want to be; he is showing disrespect for our group and its ways. There are then two reasons not to steal food from the feast: one that is second-personal (to show concern and respect for others) and one that is conventional and built on top of this (to conform to our norms). This way of looking at things is similar to that of Nichols (2004), who has focused on the question of why contemporary individuals consider violations of moral norms (he mainly focuses on those concerning harm) to be more serious than violations of conventional norms. Nichols posits that moral norms are those that target actions toward which humans already feel some emotional attraction or repulsion independently (e.g., repulsion at seeing someone assaulted); they are thus "norms with feeling." While agreeing in spirit, our claim would be that this focus is too narrow. To provide a full account of the difference between more conventional and more moral norms, one must focus more broadly on the second-personal morality involved as the underlying generator of the relevant emotions. These include, in addition to a sense of sympathy for a person harmed, a sense of fairness for a person treated with disrespect, generating an emotion of resentment or indignation. Cultural norms dealing with issues of fairness and respect thus, and for this very reason, take on a moral dimension as well.

Importantly, once norm-based expectations are begun, they cover almost everything, including even sympathy-based acts such as helping. Thus, if I ask you to fetch me a spear lying right next to you, we all expect you to do it, but if I ask you to fetch me a spear from fifty miles away, my request is out of the norm, so there are no such expectations. But again, the moral dimension does not come from the social norm per se, which only sets parameters for the expectations; it comes from the underlying second-personal morality of sympathy and harm, fairness and unfairness, in which the norm is grounded. Conventions by themselves just mandate conformity, and this includes both trivial things and things of great moral moment. But it is also possible for mere conventions to become moralized into moral norms, for example, as it becomes common-ground knowledge that a certain action has begun evoking in almost everyone in the group a second-personal moral evaluation of a uniform type. Thus, if everyone comes to think that wearing shabby clothes to a feast shows disrespect for the chief, and they resent it, then what was previously only a conventional norm becomes moralized. In general, merely conventional behaviors are moralized as individuals come to share in cultural common ground morally evaluative attitudes about particular types of social actions.[3]

Modern human individuals thus conformed to social norms for at least three immediately prudential reasons: to make sure others could identify them as in-group members, to coordinate with the group, and to avoid punishment, including threats to reputation (Bicchieri, 2006). And for modern humans threats to reputation were now amplified geometrically via gossip in a conventional language such that getting caught cheating by a single individual on a single occasion could be disastrous since everyone would soon know about it. Now the threat of gossip was to one's fully public reputation, and in a cultural world public reputation is everything. This effect is so strong that simply being watched keeps people mostly in line most of the time, as much research with contemporary humans in many different settings attests: (1) people contribute more in a public goods game if their reputation is at stake (Rockenbach and Milinski, 2006); (2) "tragedies of the common" are ameliorated if the participants know that their reputation is at stake (Milinski et al., 2002); (3) people prefer to play economic games with unknown others with, rather than without, reputational information and punishment (Guererk et al., 2006); (4) on eBay and other such websites, people become more cooperative if reputational information is an integral part of the process (Resnick et al., 2006); and (5) people behave more cooperatively even if they see a picture of eyes, rather than some other picture, on the wall in front of them (Haley and Fessler, 2005).

But modern humans did not just relate to social norms prudentially. The key observation is that, in addition to following social norms, modern humans also enforce social norms on one another, including from a third-party position as an unaffected observer. Such third-party punishment would seem to be coming from motivations less immediately and directly prudential. Thus, empirical studies have demonstrated that from as young as three years of age, children will intervene to sanction others for social norm violations on behalf of third parties (whereas great apes do not engage in third-party punishment at all [Riedl et al., 2012]). For example, if an individual attempts to appropriate or destroy someone else's property, young children will intervene to stop the transgression (Rossano et al., 2011; Vaish et al., 2011b). And children of this same age will even intervene to protect the entitlements of others, for example, when an actor has been entitled by an authority to do X, and then another person attempts to prevent her from doing it, children stand up for her rights, in essence enforcing against the attempted enforcer (Schmidt et al., 2013).

Although the content of social norms may vary more or less widely across cultures, the processes of creating, following, and enforcing social norms is almost certainly a cultural universal. Marlowe and Berbesque (2008) present evidence that some contemporary hunter-gatherers rarely engage in norm enforcement or third-party punishment of any type; they mostly just move away from norm violators. But this just shows a preference for partner choice over partner control. These same individuals most certainly do punish third parties on some occasions, especially when leaving is not an option. And they punish the reputations of third parties through gossip on a regular basis, sometimes even as they are packing up their things to leave (F. W. Marlowe, personal communication). In all, it is highly unlikely that any modern human group would not have at its disposal the possibility of individuals enforcing social norms on third parties—in one way or another, including reputational gossip—as a means of social control.

Because, as argued in Chapter 3, second-personal morality does not support third-party interventions, what is being protested in such cases is at bottom a lack of conformity to the norm. This interpretation is bolstered by the observation that young children also intervene (although with less emotion) against individuals who violate mere conventions. For example, if children learn that on this table we play the game this way, and then someone plays the game the wrong way for this table, children will intervene and stop him (Rakoczy et al., 2008, 2010). Norm enforcement of this type is distinctly group-minded, as young children punish in-group members when they violate conventions more often and more severely than they do outgroup members (the so-called black sheep effect; see Schmidt et al., 2012). This is presumably because in-group members should know better, and they should care about the group's smooth functioning more than outsiders. In contrast, in the case of moral norms involving issues of sympathy and fairness, as grounded in second-personal morality, already by three years of age young children see them as applying not just to in-group members but to all humans (see Turiel, 2006, for a review).

In all of these types of third-party intervention, children quite often use generic normative language, as in "one cannot do that" or even "it's wrong to do that!" As noted above in the discussion of intentional pedagogy, such generic language suggests that the norm enforcer is acting not just as an individual expressing a personal opinion but, rather, as a kind of representative of the cultural group (i.e., she assumes Darwall's [2013] "representative authority" to intervene on behalf of the moral community). The fact that particular social norms were created by the group is prima facie evidence for the individual that they are good for the group and its functioning, and this makes it a good thing, a legitimate group-minded thing, for the individual to enforce them on everyone. Thus, even preschool children prefer to interact with individuals who enforce social norms (even though they are acting somewhat aggressively) over those who do not (Vaish et al. submitted), presumably because such enforcement signals their cultural identity with the group and its ways.

Social norms are thus presented to individuals as impartial and objective. In principle, any member of the culture may be their voice, as representative of the culture and its values. In principle, any member of the culture may be their target, since they apply in an agent-independent manner to everyone alike (perhaps within some demographic or contextual specifications). And, in principle, the standards themselves are "objective" in that they are not how the enforcer or other group members want things to be but, rather, the morally right and wrong ways for things to be. This three-way generality—agent, target, and standards—explains why the voice of norm enforcement, like that of teaching, is generic: "It is wrong to do that." The norm enforcer refers the violator to an objective world of values that he himself may consult directly and impartially to see that his behavior is morally wrong. Joyce (2006, p. 117) argues that the objectification of moral judgments in this way is crucial to the perceived legitimacy of the process because it provides a common metric by which one may judge one's own and another's moral values, "bonding individuals together in a shared justificatory structure." Thus, social norms, like cultural information conveyed by pedagogy, come to have an independent and objective reality for individuals such that moral violations become less about injuring the particular person and more about rupturing the moral order.

Cultural Institutions

Sometimes individuals enter situations in which there are no established conventions or norms, so for social control they must invent some for themselves. For example, in a recent experiment three five-year-old children were faced with a complex game apparatus and told only that the goal was to have the balls come out the end into a bucket (Göckeritz et al., 2014). In playing the game repeatedly, the children encountered certain recurrent obstacles that they had to overcome. Their reaction to these obstacles was not just to try to overcome them but, over time, to create explicit rules for how to do this. Thus, when it later came time to show naïve individuals how to play the game, the children did so using generic normative language, as in "one must do it like this" or "it works like this." It is possible that the children in this study did not think that they really created these rules but, rather, that there were some already existing rules that they had just been lucky enough to discover. However, in a follow-up study, children were just given some objects and explicitly told there were no goals or rules; they should just play as they wished. Even in this case, the children invented their own rules and continued to pass them on to others in an explicitly normative fashion (Göckeritz et al., submitted). The point is that even when children create rules for themselves, they are committed to the idea that rules for how we do things are necessary and that we should all be committed to them normatively.

The ultimate in this process of intentionally creating rules is institutions. Although there is no crystal clear dividing line between social norms and institutions, institutions are ways of doing things that are explicitly created to meet collective group goals, so they are explicitly public. For example, modern humans were presumably pair bonding and mating in accordance with informal social norms before, at some point, some societies began institutionalizing marriage by drawing up explicit sets of rules for who can marry whom, what is an appropriate dowry or bride price, where the couple should live, what happens to the children if one person abandons the marriage, and so forth. And the marriage often was performed in a public ceremony with publicly expressed commitments (aka promises). Knight (1992), among others, argues that individuals are driven to institutionalize activities when the expected benefits are being unacceptably diminished by the costs of inefficiencies, disputes, and norm enforcement (e.g., by the "transaction costs" involved in settling disputes over bride price or compensation for abandonment). Individuals thus explicitly and publicly promise to bind themselves to certain institutional rules. The advantage to individuals is that they can now better predict what others will do, and in addition, punishments are delivered impersonally by the institution or group, so that no single individual has to bear the risks and costs. Ideally, the diminution of undesirable transaction costs through institutionalization means that many problems involving public goods are alleviated, and everyone benefits.

Institutions typically comprise constitutive norms or rules that create new cultural realities. In the famous formulation of Searle (1995), new status functions, as he calls them, are defined by the formula X counts as Y in the context C; for example, Joe counts as chief in the context of group decision making. According to Searle (2010), the fact that Joe is chief is an institutional fact with the same objective status as the fact that Mount Kilimanjaro is the tallest mountain in our universe. That is to say, despite the fact that the notion of chief is culturally created—it would not exist if there were no such thing as human beings—it nevertheless is now an objective fact about the world. Importantly, institutionally created statuses come with obligations and entitlements; indeed, one might say that obligations and entitlements define the status. For example, our chief has the obligation to make sure that all disputes over resources are resolved according to our agreed-upon rules, and he is entitled to stipulate and enforce a resolution if necessary. And we are obliged to respect his entitlements. A culturally created entity (including cultural artifacts, e.g., shells for money) may thus assume a new deontic status, a new part of "objective" reality, within the group. One may also think of promises in this

social contract is thus legitimate both because it is part of my cultural identity—"we" made it for "us"—and because I cannot help but believe that all group members, including myself, can give compatriots the sympathy and respect they deserve only by upholding this contract.

Whereas early human individuals needed to create and keep their identity as a competent cooperative partner—by acting like one—now, with modern humans, the need was to create and keep an identity as a competent cultural agent, a "person," in a particular group by doing things in conventional ways, by chastising those (including the self) who do not do things in conventional ways, by collaborating with in-group members and excluding outgroup members in everything from foraging to group defense, and in general by behaving in ways that show a special concern for one's in-group compatriots and for the group as a whole. One owed this both to one's compatriots and to oneself.

Moral Self-Governance

The modern human individual was thus born into a world of preexisting cultural conventions, norms, and institutions that had an independent "objective" existence. She conformed to their strictures in order to better coordinate with her cultural compatriots and also to avoid negative evaluations from others. But in addition—and less prudentially—she identified, as a kind of coauthor, with the originators of these supraindividual social structures from a kind of person-independent and objectifying perspective: "we" created these socially self-regulating devices for "our" benefit; they reflect the right ways to do things. In some cases, this cultural identification and objectification were given additional force by the fact that the norms referred to issues that the individual already viewed from within the framework of her natural second-personal morality.

And so, whereas a pair of early humans could make with one another a joint commitment, which lived and died with their collaborative engagement, modern humans could also make more permanent collective commitments to the culture's "objectively" right ways of doing things in general. These commitments made them legitimate not only as norms to be enforced on others but also for one's own moral self-governance. Collective commitments thus transformed early humans' sense of second-personal responsibility to a partner into modern humans' broader sense of obligation to the "objective" values of the cultural group.

Collective Commitment and Guilt

Modern humans' collective commitments to their supraindividual norms and institutions created a sense of obligation to do the right thing. Acting from this sense was the culturally rational thing to do, as right and wrong are objective values that one simply must recognize. Indeed, for modern humans the sense that one ought to do the right thing could potentially override selfish motivations, including for strategic reputation management. For instance, even though young contemporary children engage in active attempts to manage the impression they are making on others (Engelmann et al., 2012; Haun and Tomasello, 2011), in some situations they can override these strategic motivations in order to act morally. For instance, in a recent study five-year-old children were told that they could either keep some reward for themselves or else donate it to a needy child. They then watched as three other children decided to keep it for themselves. Despite this temptation to conform and also to benefit themselves, many children did the right thing and donated the reward to the needy child (i.e., more than in a control condition; Engelmann et al., submitted). The natural way of thinking about the children's behavior in this study is that their sense of obligation to do the right thing overrode their own self-serving and strategic motives.

In general, we may depict the process of collective commitment to the group's ideals of right and wrong as in Figure 4.1. In contrast to Figure 3.1, the analogous diagram for early humans in Chapter 3 (for a joint commitment), the supraindividual entity that modern human used to self-regulate was not just the "we" of immediate participants but the "we" of the cultural group, in particular, as manifest in the "objective" values embodied in the group's social norms. Also in contrast to early humans, the dotted arrows going upward indicate not that the individuals created the norms themselves (though, as we have seen, they might do so in some limited contexts) but, rather, that they have in some sense affirmed them or identified with them. The solid arrows going downward indicate once again that the supraindividual social structure is self-regulating or self-governing the interactions and relationships of all relevant persons, namely, "we" in this culture who are both authors and subjects of these norms.

Importantly, modern human individuals judged others not only for their conformity to a collective commitment but also for how they judged the nonconformity of others. That is to say, if an individual breaks a moral norm by stealing food, he should be judged harshly and punished, and so someone who judges him harshly and punishes him has done a good and just thing (Gibbard, 1990). Indeed, Mameli (2013, p. 907) goes so far as to say that these meta-judgments about others' judgments are integral to a truly moral point of view: "We do not just find torture morally outrageous, we also find outrageous that some do not find it outrageous. A person who does not find outrageous that some people do not find torture outrageous strikes us as having an incomplete grasp of the moral wrongness of torture, and this is true even if this person has no inclination or desire to participate in acts of torture." Internalizing this social process—turning it on the self, so to speak—modern humans began to engage in a type of moral self-governance. Thus, just as they judged others for their moral judgments, and others judged them in this same way,

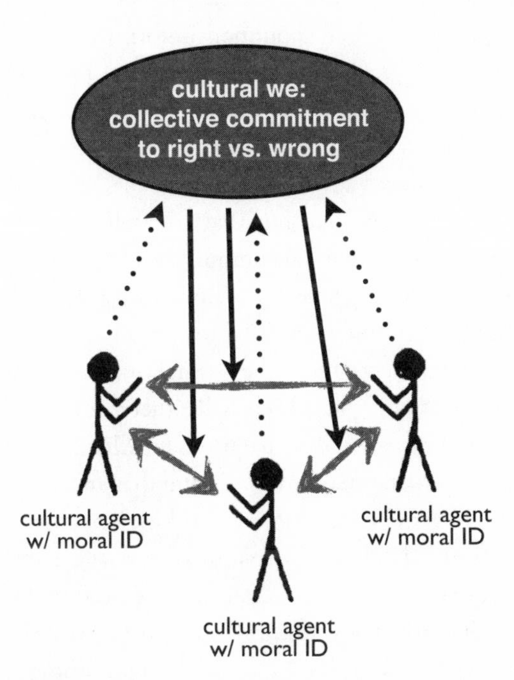

FIGURE 4.1. A collective commitment to do the right thing created or affirmed by (dotted arrows going up) and self-regulated by (solid arrows going down) cultural agents with cultural identities (ID). Cultural agents feel obligated to their compatriots to choose right over wrong ways of doing things (i.e., to follow social norms) and to make sure that others do as well (bidirectional arrows). Internalization of the process constitutes moral self-governance reflecting a group-minded cultural rationality and normativity.

they now used their skills of role reversal evaluation to judge themselves for their own moral judgments. The individual now became capable of asking herself, before acting, the self-evaluative question of whether this is a good goal to pursue or a good value to use for guidance, just as she might ask this of another's action, what Korsgaard (1996a) calls "reflective endorsement." The ability to self-reflect on and evaluate one's goals and values further rationalizes and objectifies, and so legitimates, one's commitment to moral norms over and above any tendencies toward strategic social action.

Reflective endorsement is prospective—it helps the individual decide what to do—but a similar process works retrospectively as well. In the case in which I have already done something wrong, I now judge that my previous judgment that it was the right thing to do was faulty, and so I deserve censure. Guilt of this type is thus not a fear of punishment; indeed, the feeling is that punishment is deserved. Nor is it a reaction to the violation of any and all conventions; it is not feeling bad about nonconformity per se. Rather, guilt is selectively aimed at my previous judgment of moral rightness; I thought at the time it was the right thing to do, but I no longer do. The overt response to guilt is thus to make reparations for the harm done (Tangney and Dearing, 2004), to undo the regretted act as much as possible. And it is important not just that the damage get repaired but that I myself repair it so as to bring myself back into line with my moral community and identity (as in the Vaish et al., in press study cited in Chapter 3). This contrasts with shame, in which the main issue is whether others are watching an act and whether this affects their reputational assessment of me. The normal response to shame situations is thus to withdraw and hope that others will either forgive or forget (Tangney and Dearing, 2004): I cannot make reparations for what I did because I cannot undo the information that others now have, and that affects their reputational judgments of me; I have lost face (Brown and Levinson, 1987).

The fact that guilt is a judgment about one's previous judgment comes out clearly in the fact that humans quite often feel the need to display their guilt overtly, in everything from body postures to verbal apologies. This display may preempt punishment from others—I am already punishing myself (and my suffering evokes your sympathy) so you don't need to—and this may be seen as strategic. Indeed, when viewing different individuals violating the same norm, even young children feel more positively about the one who shows guilt for having broken it than for someone who breaks it and seemingly does not care (Vaish et al., 2011a). But guilt displays also perform the even more vital function of letting everyone know (including myself) that I now acknowledge publicly that I made a bad judgment and that I regret it. I thus show solidarity with those who judge me harshly, and indeed, I agree that this negative judgment of my former judgment is deserved and legitimate. Feeling guilty for violating a moral norm thus goes beyond a strategic concern for self-reputation, and even beyond simple regret for what happened; it is a negative judgment, using the group's "objective" standards, about my previous poor judgment.

Reflective endorsement and guilt, therefore, represent a new kind of social self-regulation, an internalized and reflective self-regulation comprising multiple levels of moral judgment. And the judgments in these cases are moral because they are accompanied by a sense that they are deserved. I ought to make this judgment; it is part of my moral identity. Of course, some individuals in our group might follow our conventions, norms, and institutions strategically, always with an eye to self-interest. But these sociopaths are not moral persons and so cannot be fully trusted. "We" are those who genuinely believe—by virtue of our cultural and moral identities—that there are certain things that persons in our moral community owe to one another.

Moral Identity

As a general summary statement, we may say that what most clearly distinguishes modern human moral self-governance from any form of strategic cooperation is the role of the individual's sense of moral identity. The proximate psychological mechanisms responsible for human moral action do not involve, essentially, prudential concerns for one's self-serving interests or strategic calculations of one's own reputation; they involve moral judgments by a moral self (with the representative authority of the moral community) that endures over time and that judges the self impartially in the same way that it judges others (Blasi, 1984; Hardy and Carlo, 2005). The process is captured informally by locutions that people often produce of the type, "I could never do that," "I couldn't live with myself if I did that," "that is just not me," and so forth.

The formative process, as we are envisioning it, would go something like this: Starting as children, individuals are constantly making decisions that affect other persons, and they judge others for their actions that affect others as well. They also make judgments, and are judged by others, about moral judgments themselves. All of these judgments are coming from individuals understood as representative of the culture or moral community, as representative of the ideals of right and wrong that "we" have created over historical time. And thus the developing child begins to judge himself from this same perspective: "we" judging "me." He internalizes the process and so begins to form a moral identity that stipulates how he should act to continue being who he is. At the core of this moral identity are four sets of concerns (see inner circle of Figure 4.2): *me-concerns,* my self-interested motives aimed at helping me to survive and thrive; *you-concerns,* expressed in sympathy and helping toward others and the group; *equality-concerns,* in which others and the self are seen as equally deserving individuals; and *we-concerns,* both those emanating from a dyadic "we" formed in face-to-face interaction with a second-personal agent and those emanating from a group-minded "we" formed as one comes to identify with one's cultural group.

Many moral situations in the real world contain complex combinations of many or all of these concerns, sometimes creating moral dilemmas. But in their idealized "pure" forms, each of the sets of other-regarding concerns is associated with distinct emotions. Prototypically, violations of equality and

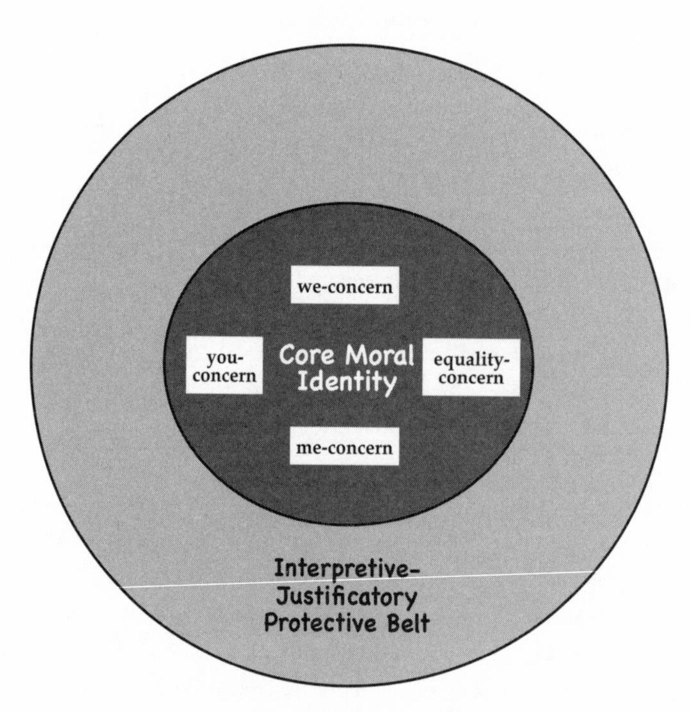

FIGURE 4.2. A moral identity model for human moral decision making.

respect are met with resentment (Strawson, 1962; Darwall, 2006): the disrespected person feels that she does not deserve to be treated in this manner and resents the perpetrator. (Some theorists think that the third-person, cultural, version of this emotion is indignation on behalf of others or the group.) In contrast, when one does not receive the sympathy that one expects, especially from a friend or other close relation, one does not feel resentment but, rather, "hurt feelings": the offended friend feels that the offender has neglected the sympathy and trust on which their relationship is built. Violations of norms or rules are often coincident with other moral violations in which someone is hurt or disrespected, but in the pure case one would simply feel a sense of disapprobation or disapproval of the rule breaker: he is not a member of the moral community because he does not follow the rules for social conduct upon which we all have agreed.

Moral decisions are typically those that involve one of the three concerns other than me-concern, even if in the end one decides that the me-concern should win out. There is thus always more or less complexity in human moral decision making. Nevertheless, the claim is that human individuals are strongly motivated to preserve their core moral identity as established by their past moral decision making. They preserve it, first of all, by acting consistently with it. But every situation is particular to some degree, and as a given situation is being assimilated to past experience its particularities must be accommodated in some way. One can thus interpret a situation of wearing a plaid coat to a funeral as being merely a breach of etiquette or as disrespecting the mourners. One can divide up resources so that the people who weigh less get less (which means, e.g., that women and children get less) under the interpretation that this preserves equality at the level of ounce-for-ounce, or one can see this as a violation of the equality of persons. And on and on. Just as scientists hold on to their core theoretical beliefs by interpreting empirical evidence in particular ways (Lakatos and Musgrave, 1970), individuals may maintain a sense of core moral identity, despite committing what others believe are immoral acts, by interpreting the situation creatively.

But the creativity of interpretation has its limits. Since one's moral identity is socially constructed, one must always be prepared to justify—both to others and to oneself—why one chose one course of action over another. Justification means showing that my actions actually emanated from values that we all share. For example, if I leave bones in the camp, and they attract scavengers, I might explain to everyone that I had to drop my butchering unexpectedly to go help a drowning child. This justification is likely to be accepted because we all accept together that helping a drowning child is more important than following the cleaning-up norm. But if I attempt to justify my negligence by pleading I needed a nap because I partied last night, this is not likely to be accepted. Haidt (2012) and others have argued that moral actions emanate from intuitive and emotional sources and that verbal justifications such as these are post hoc rationalizations aimed only at convincing others. This may be true, but this convincing of others is not just strategic, to keep others from punishing me; it is also necessary for keeping my own cultural and moral identities intact. Indeed, in some contractualist theories, the rational basis of morality lies specifically in the shared justificatory structures on which the individuals of a moral community rely (Scanlon, 1998). In creating new conventions and norms, or in choosing one norm over another, or in ignoring a norm, an individual must be prepared to justify—to give reasons for—her choices both to others and to herself in ways that ground her actions in the shared values of the moral community. The point is that although moral justification is not necessarily aimed at producing moral actions or achieving empirical accuracy, it is aimed at finding shared values that demonstrate one's continued identification with the moral community.[5]

There thus comes to surround one's core moral identity a protective "belt" of interpretations and justifications (see outer circle in Figure 4.2). Perhaps I did not share all the honey I gathered with others as I should have, but I am ill and need more nutrition than them at the moment. Does this justify my behavior to others and to myself? I practice apartheid in my society, but this is justified because God willed it so. Does this justify my behavior to others and to myself? If the answer to these questions is yes, then one proceeds ahead in the status quo. But if the answer is no, then my core moral identity is challenged, and I must do something to repair it, for example, display guilt, apologize, or make reparations. Also crucial to the process, of course, is whether an affected person is or is not a member of one's moral community—slavery could not have been in any way justified if the slaves were seen as members of the moral community—and who is the reference group for one's justifications (in this case, the slaves, the slave holders, or others). Justification is thus still another means by which the individual identifies with his group and its shared ways of life, and so legitimates his participation in his culture's social contract.

Finally, critical to the whole process of moral self-governance via moral identity is the recognition that one is free to go beyond the culture's social norms if necessary—and indeed, this freedom makes the force of obligation all the more binding because then one owns one's decisions, as it were (Kant, 1785/1988). In particular, solving moral dilemmas involving conflicting norms requires a personal weighing of values in a manner that often conforms to no conventional pattern. Both Mead (1934) and Bergson (1935) have stressed that moral beings can never escape the feeling of obligation to make principled decisions when the norms of their group do not apply in a straightforward manner or, more problematically, when they conflict with one another. This may be especially true in the complexities of contemporary multicultural societies, but even in simpler and more homogeneous societies one's second-personal morality, or one's general great ape preference for kin and friends, may in many cases conflict with aspects of one's norm-based cultural morality. Should I steal or harm others to save my friend or a compatriot in need? Thus, individuals must always in some sense freely assent to and identify with the moral decisions that they make, conventional or more personally creative. The modern human individual's life thus became, as Sellars (1963) would have it, "fraught with ought."

Distributive Justice

An especially important problem in modern human societies with a division of labor is of course the distribution of resources among persons. Indeed, how resources are distributed among persons in a society is in many ways the most concrete manifestation of its particular sense of justice (Rawls, 1971). In Chapter 3 we argued and presented evidence that young children—and, by hypothesis, early humans—have a strong tendency to share the spoils of a collaborative effort equally among partners. If modern humans now view themselves as part of one big collaborative group, then we might expect them to share resources equally among group members, even those who have not participated in procuring them, and without any special privileging of the self.

This is mostly true. The key natural phenomenon is central-place foraging, characteristic of modern human cultural groups, in which some individuals obtain food and bring it back to a central location to share with their compatriots. Observations of contemporary foragers reveal many complexities of this process, with many different ways that different groups normativize how, when, and with whom food is shared (Gurven, 2004). But one thing is common to all cultures: everyone in the group deserves food. Thus, all forager cultures have lazy or incapacitated members who contribute little, and these individuals are disadvantaged in many ways. But there is still a sense that they are one of us and so cannot be left to starve. This feature of modern human forager societies is strong evidence that, in addition to a sense of equal deservingness among partners in a collaboration, modern humans also had a sense of a minimum deservingness among all who are a part of our cultural group.[6]

Similarly, in Western industrial societies school-age children think that everyone in the group—not just collaborators—is equally deserving of a share of resources. For example, Fehr et al. (2008) found that Swiss children from the age of seven or eight years prefer an equal sharing of windfall resources between themselves and another child, and indeed, they even sacrifice resources themselves to make sure that there is an equal division (see also Olson and Spelke, 2008; Blake and McAuliffe, 2011; Smith et al., 2013), provided the partner is an in-group member. In another experimental paradigm, five-year-old children will punish anyone who does not share with them equally (i.e., in a mini-ultimatum game; Wittig et al., 2013), which chimpanzees do not do (Jensen et al., 2007). And in some cases school-age children who are dividing up resources will go so far as to throw away extra resources rather than allocate them in a way that violates equality (Shaw and Olson, 2012). This developmental pattern suggests that collaboration may have been the natural home for an equality preference in early human evolution, but with the emergence of cultural groups and a sense of the entire group as a single collaborative enterprise, everyone in the group was seen as a deserving recipient of resources, especially large-packet resources, collected by any group member.

Recently a number of cross-cultural studies—most of them including both small-scale and Western industrialized cultures—have found cultural differences in the way that children choose to distribute resources. For example, Rochat et al. (2009) found cultural differences in how three- and five-year-old children chose to divide a set of windfall resources between themselves and various other individuals. House et al. (2013) found that young children from six different societies behaved similarly in a dictator game requiring them to sacrifice resources, until they were about eight or nine years of age, at which point cultural differences emerged. Zeidler et al. (in press) found significant cross-cultural differences in children's willingness to take turns accessing a monopolizable resource. Henrich et al. (2001) established clear differences in how adults in different cultures behave in an ultimatum game, though there was some sense of fairness in all of the fifteen small-scale societies studied. Finally, Schäfer et al. (in press) found that school-age children from a Western industrialized society divided resources among collaborative partners proportionally to the work productivity of each, whereas children of the same age from two small-scale African societies did not.

The key to understanding these cultural differences in distributing resources is the ontogenetic pattern. Younger preschool children from different cultures (who actually are not well represented in any of the cited studies) differ little in their sense of fairness because they are all operating with a natural, second-personal morality. But later, especially during school age, children begin subscribing to the social norms that their culture has worked out for distributing resources in fair ways. In general, in the current view, individuals of all ages in all cultures operate with multiple motives that come into play when resources are being distributed: selfish motives, other-directed sympathetic motives, and motives of fairness based on deservingness. But exigencies in the lifeways of different cultures have necessitated that these different motives be combined and weighted in different ways in their respective social norms, and this is not just between cultures but sometimes in different situations within the same culture. Thus, for example, in situations of extreme scarcity, people in all cultures are likely to expect and tolerate a certain amount of selfishness. In situations such as the ultimatum game, people in all societies have at least some sense of fairness. And even though children from some small-scale cultures do not take work productivity so much into account, almost certainly the children and adults in those cultures believe that some people merit more resources than others (e.g., the chief over others, or adults over children). We would therefore claim that even in the most basic domain of distributive justice, as well as other social domains, all normally functioning human beings possess a universal second-personal morality, with a cultural morality of social norms layered on top.

Finally, we must stress again that the notion of distributive justice is not fundamentally about "the stuff" but, rather, about being treated fairly and with respect (Honneth, 1995). When there is no way to divide resources in a fair manner, contemporary adults and children will accept almost any allocation of resources as long as a fair procedure has been followed. Thus, both children and adults respect even an unequal division of resources if it results from, for example, casting lots, drawing straws, rolling dice, or playing rock-paper-scissors. In two recent studies, young children were satisfied to receive a less than equal portion of a resource if the division had been determined by a randomized "wheel of fortune" (Shaw and Olson, 2014; Grocke et al., in press). In procedural justice, as this process is known, rules are formulated impartially, that is, without knowing ahead of time how particular individuals will be affected (under a "veil of ignorance" [Rawls, 1971]), and as long as individuals know that this is the process, they are satisfied with the outcome. They are being treated justly and with due respect, and that is what is important. Indeed, the institution of private property, as practiced by many contemporary cultures, may be seen as essentially comprising the codification of respect between individuals with regard to objects. It works because, and only because, individuals respect the ownership rights of others. Consistent with this view, Rossano et al. (2015) found that five-year-olds, but not three-year-olds, showed a respect for the property rights that others had signaled by doing such things as piling toys into a corner to assert ownership.

Cultural Group Selection

For certain, human beings in general have some built-in responses to morally relevant situations based on intuition and emotion that have evolved to deal with evolutionarily important situations, especially those in which there is no time for a considered decision (see Haidt, 2012). In addition, we would claim, in their more considered moral decision making humans in general have the four sets of concerns outlined above (see Figure 4.2). But over and above these human universals, particular cultures have created their own social norms and institutions that bundle together various aspects of these concerns in novel ways and place them, as it were, under a very strong we-concern: if you identify with our cultural group and its social norms, when this kind of situation arises you must follow the norm. The culture creates such norms to deal with recurrent situations that arise in their particular lifeways: they reinforce individuals' natural morality with rules that provide independent additional reasons for "doing the right thing." Thus, while some cultures have to deal with a scarcity of resources, others do not; while some cultures have to deal with intense competition from neighbors, others do not; while for some cultures group decisions about where to live for the next few weeks are a matter of life and death, for others life is more sedentary; while in some cultures individuals accumulate capital and so have power over others, in others this does not apply; and on and on. Cultures build on top of individuals' natural morality a cultural morality encouraging conformity to norms designed for maintaining social order in particular ways of life.

As noted in Chapter 2, with the emergence of modern humans—or, perhaps, only later as the demographic conditions of group competition emerged with greater intensity—there came a new process in which such cultural variability played a key role. That new process was cultural group selection. Perhaps from before they began spreading out of Africa, some one hundred thousand years ago, different cultural groups began creating different conventions, norms, and institutions, for all of the ecological reasons listed above and more. All of these different supraindividual social structures aimed to coordinate interactions among individuals and to cooperativize potential conflicts within their respective groups, but they did so in different specific ways. Because compatriots within a culture insisted that everyone conform to their group's norms, they persisted stably within a group over historical time. This stability produced maximal within-group homogeneity, and this in turn, when combined with substantial between-group heterogeneity, led to systematic variation in norms among different cultures. The result was a new process of cultural evolution.

Because these group-specific behaviors and structures were differentially effective in regulating group life, those modern human groups with the conventions, norms, and institutions that best promoted cooperation and group cohesion won out, by either eliminating or assimilating competitors from other groups (Richerson and Boyd, 2005). This meant that there now existed an evolutionary process on the level of the cultural group, and it played an important additional role in the nature of modern human group-minded morality, as members of cultures socially selected individuals with whom they wished to share their cultural lives. And so, in one way or another, each group created its own specific instruments of social control that coordinated group activities and encouraged cooperation in their particular environment—their particular practices, norms, and institutions—and those that did the best job promoted the survival of the group that created them.

Cultural groups thus came to constitute a new form of plural agency. Analogous to early humans' joint agency in collaborative dyads, modern humans created a kind of collective or cultural agency in which, in one way or another, there had to be group decisions about where we travel, how we deal with neighboring groups, how we construct our temporary camps, and so on. If an intentional agent is a living being who not only acts intentionally toward a goal but also knows what she is doing and so self-regulates the process as unexpected contingencies arise, then a cultural group acting toward a collective group goal, with collective commitments to self-regulate their progress toward that collective goal, may be seen as a collective agent (List and Pettit, 2011).

BOX 2. Preschool Children's Nascent Norm-Based Morality

Preschool children, prototypically three to five years of age, are not fully moral beings. Nevertheless, they show some new and morally relevant behaviors and judgments relative to younger toddlers, many concerning group-minded things, such as group identity and social norms. This fact is at least potentially relevant to the evolutionary proposal that early humans' second-personal morality was followed by a more group-minded, norm-based morality. Here is a summary of the most apposite studies as reported in this chapter. In all cases, children were studied either at five years of age only or at ages three and five (in many cases the procedures were not tried with younger children because the researchers supposed that they would not work with them). Research on children's in-group bias is summarized at the beginning of this chapter.

Collective intentionality: Preschool children but likely not toddlers

- understand that everyone from their in-group should know certain things in cultural common ground, even if they themselves have not seen those others actually experiencing them (Liebal et al., 2013);
- feel responsible for harmful acts committed by others from their in-group but not by those from an outgroup (Over et al., submitted);
- have a tendency to divide resources equally, even paying a cost to reject unequal divisions (e.g., in an ultimatum game; Wittig et al., 2013);
- judge that a division of resources is fair even if they themselves end up with a lesser amount, provided that the procedure for making the division was impartial (Shaw and Olson., 2014; Grocke et al., in press); and
- respect the ownership rights of others (Rossano et al., 2015).

Social norms: Preschool children but likely not toddlers

- intentionally punish others who harm third parties (McAuliffe et al., 2015; Riedl et al., 2015), which chimpanzees do not do (Riedl et al., 2012);
- verbally enforce both moral and conventional norms on third parties (see Schmidt and Tomasello, 2012, for a review of the many studies);
- prefer individuals who enforce social norms to those who do not (Vaish et al., submitted); and

- tend to teach and enforce norms with generic normative language from an objective stance (e.g., "that is wrong"; Köymen et al., 2014, in press).

Collective commitment and obligation: Preschool children but likely not toddlers

- avoid conflicts by creating coordinating conventions for all in the group (Göckeritz et al., 2014),
- resist peer pressure in order to "do the right thing" in some circumstances (Engelmann et al., submitted),
- feel guilty only for harm caused by them (or an in-group compatriot) and feel a special responsibility to repair such harm (Vaish et al., in press), and
- prefer individuals who show guilt for their transgressions over those who do not (Vaish et al., 2011a).

The Original Right and Wrong

Our attempt in this chapter has been to conceptualize modern human moral psychology—built on top of early human second-personal morality—as a set of adaptations for living in larger, tribally organized social groups, aka cultures. As modern human cultural groups entered into a variety of new ecological niches, each created its own special set of conventionalized cultural practices, norms, and institutions adapted for local conditions.

To survive and thrive in this new cultural world, what individuals had to do most urgently was to conform to the cultural practices of the group. They conformed, in the first place, for Hobbesian prudential reasons: to affiliate with others in the group, to coordinate with others in the group (including strangers), and to avoid punishment for nonconformity (including reputational gossip). They conformed also for Rousseauean "legitimacy" reasons: although they had not themselves created the social contracts into which they were born, they identified with the authors and recognized as valid the "objective" values of right and wrong that they represented, thereby legitimating them.

In addition to being legitimate, a subset of the culture's norms were also seen as moral, based on members' already existing second-personal morality. In classic social contract narratives, the isolated individuals who are contemplating joining forces are in no way moral, natural or otherwise. But that is precisely the point of our two-step story. In our story, from the beginning modern humans possessed early humans' second-personal morality for face-to-face interactions with collaborative partners, so they did not have to create an "objective" morality from scratch, only scale up their existing second-personal morality to fit a cultural way of life.

Modern Human Moral Psychology

This "scaling-up" account of the transition from early human to modern human moral psychology specifies transformations in the four sets of psychological processes most relevant to human morality: (1) sympathetic concern for partner welfare transformed into loyalty to the group, (2) cognitive processes of joint intentionality transformed into collective intentionality and its resultant objectification of values, (3) social-interactional processes of second-personal agency and cooperative identity transformed into cultural agency and moral identity, and (4) self-regulatory processes of joint commitment and responsibility transformed into moral self-governance and obligation. Analogous to the claim about early human second-personal morality, the claim for modern humans is that many elements in these transitions were adaptations not for morality per se but, rather, for coordinating with others cognitively. In combination with various kinds of cooperative motives, this created a distinctively new moral psychology.

IDENTIFICATION AND LOYALTY. Early humans felt sympathy for their interdependent collaborative partners. In becoming culturally organized, modern human groups became interdependent collaborative entities themselves (competing with other groups), with which their members identified by conforming to their ways of doing things. Members of cultures thus had sympathy for all of their compatriots with whom they were interdependent and for the group as such, which they did not have for any outgroup barbarians in the vicinity. They were loyal to the group and genuinely valued its smooth functioning. This loyalty was important for the individual's group identity and acceptance, and so for survival.

OBJECTIFICATION. To coordinate collaborative activities with in-group strangers, as well as to coordinate large-group collaborative activities, modern human individuals scaled up their skills of joint intentionality, based on the personal common ground of collaborating partners, into skills of collective intentionality based on the cultural common ground of all in the group. These new cognitive skills enabled the creation of conventional cultural practices and institutions, the different roles in which everyone knew that everyone knew how to perform in the ideal way. Because individuals understood that "anyone" could plug into these roles, there emerged a sense of the agent independence, and thus "objectivity" of the role ideals. Everyone thus knew in cultural common ground the correct and incorrect ways that various roles needed to be played for group success, and this included the most general role of simply being a contributing group member.

In this process, the early human ability to shift perspectives with a partner transformed into the ability to take the perspective of any rational being. Modern human individuals could thus take a fully "objective" view from nowhere on situations—the view that any rational person would take—which enabled each to act as a fully "impartial spectator" in situations requiring a balancing of perspectives among many diverse parties, including the self. As was the case with early humans' sense of self–other equivalence, this agent-independent view from nowhere was not on its own a moral judgment, merely the recognition of how things were in this new social reality. Children in modern human cultures were taught how things are and how one should act by someone employing the generic and authoritative voice of pedagogy emanating not from that person's opinions or perspectives but, rather, from the culture at large. The group's conventional ways of doing things were thereby objectified into the right ways of doing things. This objectification was further reinforced by cultural institutions, which created new cultural entities (e.g., chiefs and boundaries) with new deontic powers of entitlement (e.g., to declare war and to define territories). Individuals were born into this institutional world, always already there, as a part of objective reality.

LEGITIMIZATION. Whereas early human collaborative partners needed some means of partner control, modern humans living in cultures needed a more expansive form of social control. The outcome was that the role-specific behavioral ideals from early humans' collaborative activities were scaled up into social norms that were administered by third parties. Social norms were created and maintained in the cultural common ground of the group—so that no one who grew up in the group could plausibly deny knowing them—and demanded conformity more or less equally from everyone. Since social norms came into existence for individuals who already had a second-personal morality, those norms that reinforced the attitudes and motivations of second-personal morality were thereby moral. Violating these moral norms could be fatal because it meant, in the extreme, not just being rejected by a partner, which could be overcome by finding a new partner, but being ostracized from the group entirely (the group makes a partner choice). In modern human cultures, the total dependence of the individual on the group—and the individual's genuine concern for the group and its smooth functioning—meant that it was culturally rational to conform.

But, moreover, violating moral norms was also simply wrong, and any punishment one received was deserved. A cooperative cultural agent recognized that everyone must conform to the group's way of doing things—and must make sure that everyone else conforms as well—or else things might very well fall apart. The fact that individuals normatively expected others to enforce the culture's norms is telling. The individual was in fact making a moral judgment about the other's moral judgment—for example, that he should have chastised her for doing that—suggesting that social norms were not conceived as something that "they" enforce on "me" and to which I should conform strategically; rather, they were legitimate guidance for the way that "we" ought to do things, the right way of doing things. Because the individual felt himself, through the process of group identification, to be in some sense a co-author of the social norms, they took on an added legitimacy, such that violating them became, in a sense, violating one's cultural identity. Therefore, to do something selfish in full knowledge of the governing social norms was to choose one's selfish self over one's cultural self, who was striving to be a "person" who did the right thing virtuously within her moral community, and so this selfish act deserved to be condemned.

MORALIZATION. Unlike early humans who made their own joint commitments with others, then, modern humans had the largest of their collective commitments already made for them, as it were, in the form of the cultural conventions, norms, and institutions into which they were born. Theirs was thus the problem of the social contract: how, when, and whether to sign onto, to legitimate from one's own personal values, these supraindividual cultural structures and their supposed group-minded rationality. In accepting coauthorship of these collective commitments, and affirming that their proper enforcement was legitimate because deserved, an individual began to create a moral identity. The normative judgments of deservingness that constituted an individual's core moral identity reflected concerns of sympathy ("you > me"), concerns of justice ("you = me"), and concerns of cultural rationality ("we > me"), whose essence was inherited from early humans' second-personal morality. To maintain one's core moral identity intact, one had to interpret situations creatively (e.g., in determining what constituted appropriate sympathy and fairness in particular situations) and, further, to justify any deviant action to others and oneself by interpreting it in terms of values that were widely shared in the moral community.

Modern human individuals internalized the self-regulating processes of collective commitment to social norms—and the values of right and wrong that these embodied—and this resulted in a sense of obligation. The sense of obligation, and its use by individuals to self-regulate their behavior in the direction of right (and not wrong) actions, constituted a kind of moral self-governance, which relied on a personal sense of moral identity for guidance. Moral self-governance meant performing actions only after reflectively endorsing them as actions that would be positively judged by anyone, and feeling guilty if one later judged that one had used poor judgment. But of course, the individual always had the option of going against the group's norms, even being immoral, and, moreover, moral demands often conflicted with one another such that no available conventional solutions were of any help in making a behavioral decision. In such cases the individual had no choice but to decide for herself which of the various forces at play should be decisive in her decision making. The moral decisions of modern human individuals were thus always populated by many different "voices," and there existed no one but the individual agent to adjudicate among them.

Overall, in an environment of social interdependence, it is rational to cooperate: one should invest in the social resources on which one depends. For modern human individuals these social resources included not just their personal relationships with others but also the cultural practices, norms, and institutions necessary for coordinating with others and maintaining a reasonable level of social control in the group. But to get to morality we need to go beyond individual rationality. We need a cooperative rationality and, in the case of modern humans, a fully cultural rationality. Individuals interact cooperatively with others not only because it is the strategic thing to do, although it is often that, but also because it is the right thing to do. It is the right thing to do because one's groupmates are in all important respects equivalent to oneself, so they deserve one's cooperation. It is also the right thing to do because the cultural practices, norms, and institutions within which we all live were created by "us" for "us" (given that we culturally identify with our forebears in the culture), and we therefore affirm their legitimacy from the outset—from a kind of veil of ignorance in which all of us are potential violators and enforcers—so that we see enforcement for violations, including against ourselves, as deserved. And it is the right thing to do because . . . well . . . it just is the right thing to do. Despite temptations to cheat or to resist enforcement against themselves, which they might do on occasion, the fact is that modern human individuals judged themselves as they judged others, and so to be uncooperative toward others in these ways would be an abdication of their personal moral identity.

Multiple Moralities

For many moral philosophers, what we have now is a principled morality, a fully objective morality of right and wrong. But let us make sure we understand clearly the different contributions of natural (second-personal) and cultural ("objective") morality. By themselves, social norms are not moral. Many conventional norms have nothing to do with morality directly. But they may, in the appropriate circumstances, be moralized, and they are moralized precisely by being connected to natural morality. Thus, one may construe dressing in rags for a celebratory feast as morally wrong if it actually does cause harm or disrespects others in the sense of ruining the celebration for them or making them feel like they are being treated as less than equals. It is thus not the nonconformity per se but, rather, the harm that nonconformity may cause or the disrespect it may show that evokes negative moral judgments.

Along these same lines, it is important that many social norms considered as moral by their practitioners are considered immoral by members of other cultures. This is due in most cases to different perceptions of what the dictates of natural morality require in particular circumstances. In the recent past in the contemporary world, for instance, apartheid was a system of social norms that practitioners believed were moral, but only with some creative definitions of harm and disrespect (or lack thereof), and some creative accounting of who was in the moral community. The practitioners at some point came to see these social norms as immoral, as they—due to various internal and external forces—came to see them as violating their own natural senses of sympathy and fairness and their sense of who was in the moral community. And so, again, it is not the norms themselves but, rather, their connection with humans' natural second-personal morality that is the deciding factor. Indeed, when outsiders to a culture perceive some of its social norms as immoral—or when insiders have a change of heart in this direction—they espouse the opinion that truly moral individuals should oppose these norms. Cultural norms do not create morality, only collectivize and objectify it, and institutions may go a step further and sacralize it.

The point is that cultural norms concern morality exactly to the degree that they connect in some way with humans' natural attitudes of sympathy and fairness that have existed since before there were group-minded, norm-based cultural groups. Following social norms is simply conformity, and enforcing them is simply enforcing conformity through one or another form of punishment. And in another of those reflective twists that seems so characteristic of many uniquely human phenomena, violators of moral norms punish themselves as well through feelings of guilt. They take on the perspective and attitude of the group toward the judgments that they themselves have just made. Their commitment to upholding the social norms of the group, along with their ability and tendency to view themselves as simply one not-so-special member of the group, leads to the kind of self-flagellation that only humans could invent (Nietzsche, 1887/2003).

And so the cultural sense of good and right is what "we" consider to be ways of treating others with sympathy and fairness within our cultural contexts. This cultural morality, as all culturally transmitted things, is mostly conservative. But there is cultural creation as well. As new circumstances arise, cultural groups adapt by conventionalizing new social norms and formalizing new institutions. In the contemporary world, much change is precipitated by the changing demographics of different cultural groups as they come together and split apart for various economic and political reasons. The contemporary situation is complicated by the fact that it is not always clear who is and is not a member of any particular moral community. The outcome for contemporary individuals is a complex and variegated sense of morality in which different social norms often conflict with one another. Also, quite often, the demands of their group-minded cultural morality conflict with those of their second-personal natural morality, with no fully satisfactory solution apparent. And of course, they can face conflicts of values between different cultural groups—who might even be a part of their contemporary nation-state—that also seem fundamentally intractable. But our claim, perhaps hope, would be that there are resources for resolving these moral dilemmas, by coming to common-ground agreements on (1) what does and does not constitute sympathy/harm and fairness/unfairness in particular situations, and (2) who is and is not in our moral community. This then grounds our moral discourse in the natural morality shared by all of humanity.

In all, it is important to recognize the complexity and perhaps even unavoidable contradictions that reside within human morality. Its multiple sources and layers can never be applied consistently in all situations, given the messiness and unpredictability of human social life. Having sympathy for my hungry cooperative partner might lead me to allow him more than half of the spoils, but that contradicts my general tendency to divide them equally. There might be a social norm that I should not steal food from others, but what if my child or friend is starving? And what about situations where different social norms could be applied equally well? Human morality is not a monolith but a motley, patched together from a variety of different sources, under different ecological pressures, at different periods during the several million years of human evolution (Sinnott-Armstrong and Wheatley, 2012). Human beings today thus enter into each and every social interaction with selfish me-motives, sympathetic you-motives, egalitarian motives, group-minded we-motives, and a tendency to follow whatever cultural norms are in effect. In situations of deprivation, most of us would be selfish. When someone else is in dire need, most of us would be generous. In situations of equal collaboration, most of us would be egalitarian. And if we are playing one another in the finals of Wimbledon, it does not matter who needs the prize money more or who has worked the hardest for it, because the cultural norm is that whoever plays the best tennis wins the prize. All of these motives are always in some sense already there; the only question is which one, or ones, will win the day in particular situations.

Coda: After the Garden of Eden

By the time they began their diaspora out of Africa in significant numbers, beginning around one hundred thousand years ago, modern humans were thus three-ways moral. First, they had a special sympathy for kin, friends, and cooperative partners, along with a feeling of loyalty to their cultural compatriots, which motivated preferential treatment for these special people. They had a morality of sympathy. Second, they felt a responsibility to act respectfully in their direct, dyadic interactions with deserving others, which led them to treat those others fairly. They had a second-personal morality of fairness. And third, laid on top of these, as it were, they felt an obligation to the group and to themselves to conform, and to ensure that others conformed, to the impartially formulated conventions, norms, and institutions of their cultural group, especially those relating to second-personal morality. They had a group-minded cultural morality of justice.

Within the constraints of these three universally human moralities, the specific cultural moralities of modern human groups could differ from one another significantly. This meant that another important part of the process was cultural group selection, in which those cultural groups with the most cooperative and effective conventions, norms, and institutions eliminated or assimilated other competing groups. Especially important in this process were events beginning around twelve thousand years ago. It was at this point that some modern human groups stopped chasing after food and began bringing food to themselves by domesticating various plant and animal species. The rise of agriculture and the cities it spawned was of course a monumental event in the history of human sociality. When human groups began coming to the food, people with very different cultural practices—who spoke different languages, ate different foods, wore different clothes, engaged in different forms of hygiene, and so forth—all came to live in close proximity to one another. And because they were sedentary, they had to find ways to get along, despite some different social norms and values. And, of course, agriculture meant that some individuals were able to dominate a surplus of food and use it as capital to wield power over others (Marx, 1867/1977). Creating cooperative arrangements in these new social circumstances required some new supraindividual regulatory devices. Most important from the point of view of modern human morality were law and organized religion.

Formal (and at some point written) laws have classically been thought to promote human cooperation and morality mainly by adding another layer of punishment against cheaters that is administered impersonally by an institution, so that individuals do not have to take the risks or bear the costs. And this is clearly a large part of the story. But Shapiro (2011) argues that the main function of a human system of laws is not to keep mean-spirited cheaters in line but, rather, to coordinate the activities of otherwise well-meaning individuals living in large-scale cultural groups in which the goal pursuits of one individual may unintentionally thwart those of another in myriad ways. For example, in the large-scale agricultural communities of several thousand years ago, it could easily have happened that one individual promised grain to another for some tools now, but then bad weather ruined her grain harvest. What should we do to prevent a potentially destructive fight? Or it might have happened that one individual diverted water to irrigate his fields, and his action unknowingly deprived other individuals downstream of water for their crops. Or one individual traded a goat to another, and it turned out to be barren. What should we do? Shapiro (2011, p. 163) summarizes the situation:

> The disagreements that arise . . . are entirely sincere. Each of us is willing to do what we morally ought to do—the problem is that none of us knows or can agree about what that is. Customs cannot keep up with the evolving conflicts because they develop too slowly to regulate rapidly changing social conditions and are too sketchy to resolve complex disputes and coordinate large-scale social projects. While private negotiation and bargaining are able to quell some conflicts, this process can be very costly, not only in terms of time and energy but emotionally and morally as well. With many more ways to interfere with one another's pursuits and many more goods to fight over, there is a danger that disputes will proliferate and fester, causing the parties to refuse to cooperate in the next communal venture or, worse, to become involved in ongoing and entrenched feuds.

Shapiro goes on to show that what is needed is not just more social planning but a plan for social planning. For example, we may nominate a chief or a council of elders to make rules that everyone must follow and to adjudicate disputes in the application of those rules. We must now have second-order rules for how a chief or council of elders is chosen and, possibly, how they may and may not operate. Shapiro thus argues that legal systems are institutions of second-order social planning in the "circumstances of legality," which obtain whenever a community has numerous and serious moral problems whose solutions are complex, contentious, or arbitrary. And so arose legal forms of second-order social planning.

Importantly for our current concern with human morality, Shapiro (2011, p. 171) argues further that "the law aims to compensate for the deficiencies of nonlegal forms of planning by planning in the 'right' way, namely, by adopting and applying morally sensible plans in a morally legitimate manner." Laws must be "morally sensible" if those who are subject to them are to view them as morally legitimate. Of course, a dictator with weapons may formulate a rule that everyone must murder their second-born child and enforce it with military might, and one could call that a law. But this is a recipe for revolt. The laws that people follow, endorse, and even enforce on one another are ones that they can positively affirm from their own moral point of view, ones to which they are collectively committed (Gilbert, 2006). This also applies to the second-order rules for how rules are made: individuals must accept the procedural justice of those second-order (constitutional) laws to feel that they are legitimate. Therefore, what gives a system of laws legitimacy in the eyes of its subjects—and so creates in them a sense of political obligation—is that the legal point of view always *purports* to represent the moral point of view. Consequently, citizens typically follow the law and respect it as morally legitimate. That is, until they don't.

One way that leaders throughout human history have sought to legitimate themselves and their laws from a moral point of view is to claim that they have somehow been anointed by a deity or in some other supernatural way. The origins of religious attitudes in general are unknown, but one key aspect could have originated in the kind of agent-independent and group-minded thinking characteristic of modern humans, particularly when given a historical dimension, including ancestors and ancient traditions. Norms and institutions have a kind of abstract, almost supernatural existence and apply not just to particular individuals but to "anyone," however that might be conceived. A major source of wonder in human experience is where are our venerated ancestors who founded our society, and indeed, a key foundation of a religious attitude is the veneration and worship of deceased ancestors and traditions whose spirit somehow lives on (Steadman et al., 1996). Leaders then took advantage of this attitude and claimed supernatural sources for their leadership.

The organized religions that arose in association with early large-scale societies thus emerged as still further social means, in addition to legal means, for encouraging cooperation. Shared beliefs in supernatural entities and forces—accompanied by a feeling of veneration—provided additional cultural common ground, creating an even stronger in-group bond. Different religions arose within different societies many times, starting small and growing larger over time. Wilson (2002) argues that the typical pattern is for groups of poorly empowered individuals to bond together to gain power in the society by creating ever larger and more "organic" social entities that function well as a team by helping one another, supporting one another, and generally working together. So people in the same religion become more interdependent with one another. Individuals' cooperation is encouraged by many different explicit means, not the least of which is, in some cases, the omnipresence of a deity watching every move an individual makes and rewarding cooperators in a supernatural afterlife (Norenzayan, 2013). The rules one must follow are sometimes clearly aimed at encouraging cooperation (e.g., "love thy neighbor"), but they are sometimes merely ritualistic, so that in following them one simply demonstrates one's group identification. Wilson's (2002, p. 159) characterization is that "religions exist primarily for people to achieve together what they cannot achieve on their own."

Even more strongly, Durkheim (1912/2001) argues that religion springs directly from humans' communal, group-minded ways of thinking and doing things, and it does so in ways of which individuals are almost totally unaware. In communal life the sacred refers to the collective practices of the moral community—with rituals playing an important role because this is their only function—whereas the profane refers to individuals' self-interested pursuits. (Interestingly, magic, while it shares with religion some supernatural dimensions, is not religious precisely because it is utilitarian and does not unite individuals in a common life.) Durkheim thus emphasizes that religion plays a unique role in human cooperation and morality. Whereas other social institutions encourage cooperation and morality through promises of rewards and threats of punishment—and individuals may also construct a personal sense of obligation from their social and cultural interactions—religion conceptualizes cooperative and moral activities as something higher to be strived for. Individuals who are moral for religious reasons are typically not governed by a sense of obligation but, rather, by a striving for something grander. "The more sacred a moral rule becomes, the more the element of obligation tends to recede" (Durkheim, 1974, p. 70).

Religious groups, regardless of whether they are associated with political entities, invariably have a very strong sense of in-group/out-group morality (Wilson, 2002). A large part of the story of Western civilization is thus conflicts involving, in one way or another, religious groups versus "heathens." Haidt (2012) argues that this strong in-group/out-group mentality combined at some point with the human disgust response to create another important dimension of human morality: purity, or sanctity. The human disgust response is prototypically toward products that are expelled from within the human body—saliva and feces, for example—and the proposal is that this response is then metaphorically transferred to things that come from outside the group. Things that are from outside our group's common ways of doing things, especially those involving corporeality, are disgusting. Haidt further points out that humans have a tendency to moralize entities and activities as they push them outside the group's ways of doing things. For example, in contemporary life, many people who originally only mildly disliked smoking and smokers now find the entire practice disgusting. Disgust for things external to us thus provides the strongest possible contrast to the sacredness of things internal to our lifeways.

And so contemporary humans became legally and religiously moral on top of their natural and group-minded moralities. But the specific moral norms of a particular cultural group—with or without legal and religious institutions contributing—may change over historical time. Thus, within a culture there are often different "voices," some with more capital and political power and some with less. If we think of these different subgroups of a culture as analogous to a cultural group, we may then imagine a kind of cultural group selection within a population of humans as different "voices" compete to push an agenda. Kitcher (2011) eloquently describes such a process, highlighting the ways that individuals and subgroups within a culture engage in moral discourse in attempts to influence the norms, values, and institutions of the cultural group as a whole. In some cases, one subgroup has almost no "voice," for example, African-Americans or gay persons in America sixty years ago, and they remain powerless to influence the moral discourse. Harkening back to Hume's criterion that human morality and justice can emerge only among more or less equals, situations with a strong imbalance of power work against a cooperative morality (see Hume, 1751/1957).

But things can change. It can happen that so-called norm entrepreneurs take a risk and attempt to promote a value in some new way that is not already shared in the group (behavioral convergence), or else many individuals all privately have a certain attitude but it becomes shared only when it becomes mutually known among them that they all do (epistemic convergence). The problem is that people with no "voice" in the discourse almost certainly will not be seen as leaders and are not in the communicative networks of the majority. But there is one method that has been proven effective in this situation, and it is very revealing of the "natural" morality possessed by all human beings. The method, perfected by Mahatma Gandhi and Martin Luther King, is for the people without a voice to bring in front of the voiced people's eyes the way they are being treated: a kind of second-personal protest writ large. For example, in the southern United States in the 1960s, African-Americans sat at all-white lunch counters, refused to leave when asked, and so incurred the expected police brutality. The key was that it needed to be in front of many eyes, preferably television cameras. Individuals in the white majority who previously just did not, or chose not to, think about these things then had presented to them in their cultural common ground (mostly in their living rooms) exactly what was going on. African-Americans' protests did not tell the white majority precisely what to do, it only displayed their resentment, under the assumption that the majority already knew the right thing to do. At that point, some people with a voice began to take a leadership role in denouncing the immoral treatment. It worked, to the extent that it did, because it touched on some values of sympathy and fairness that the majority already possessed.

We may thus see the evolution of cultural moralities as taking place both between groups and within groups. In the millennia leading up to the advent of agriculture, cultural group selection was shaping particular human populations in the direction of greater cooperation and morality. After agriculture, multicultural civil societies have struggled to find ways to reconcile in good ways the different cultural moralities of the different subgroups within them.

5

Human Morality as Cooperation-Plus

There can be no complete moral autonomy except by cooperation.

—JEAN PIAGET, *THE MORAL JUDGMENT OF THE CHILD*

Theories of the evolution of human cooperation tend to emphasize either small-group processes in relatively early humans (e.g., Cosmides and Tooby, 2004) or large-group processes in more recent humans (e.g., Richerson and Boyd, 2005). Our view is that, for a full explanation, both of these evolutionary steps are needed (Tomasello et al., 2012). We disagree with theorists of both types, however, about the precise nature of the social and psychological processes involved. This is especially true when we dig beneath cooperation as general patterns of social interaction to the actual moral psychology at work.

At the first step, we believe that the key was not just that the groups were small—though they were, and this had a role to play—but that early humans evolved a new moral psychology for face-to-face dyadic engagement in collaborative contexts. There is much evidence that dyadic interactions have unique qualities involving such things as eye contact, voice direction, and postural adjustments during communication, such that some anthropologists have posited a human "interaction engine," geared for face-to-face dyadic interaction, as the explanation for virtually all forms of uniquely human sociality (e.g., Levinson, 2006). Moreover, a number of the most basic forms of human social interaction are fundamentally dyadic, for example, friendship, romantic love, and conversation, and the evolved emotions associated with these dyadic relationships are qualitatively distinct from anything associated with group interactions (Simmel, 1908; see Moreland, 2010, for a review of recent social psychological research). And recent philosophical analyses, as we have recounted in some detail, emphasize the many special qualities of human second-personal engagement (e.g., Darwall, 2006, 2013; Thompson, 2008).

At the second step, we believe that the key was not just that the groups were larger—though again, this was true, and it had a role to play—but that modern humans evolved a new group-minded moral psychology. Few phenomena have been better documented in anthropological research than processes of cultural identification involving "ethnic markers," such as special forms of dress, speech, and ritual, that cultural groups use to differentiate themselves from other groups (e.g., Boyd and Silk, 2009). In social psychological research, as we have reported, it has been reliably documented in myriad experimental paradigms that humans operate with a very strong in-group/out-group orientation and that large groups (especially cultural groups) play a larger role in the individual's sense of identity than do more ephemeral dyadic relationships (Fiske, 2010). And many philosophical analyses, from Hegel and Mead onward, have emphasized the way that individual human rationality and thinking are shaped by societal-level processes, such that there is even a special kind of group-minded thinking, empirical evidence for which comes from many lines of research, beginning with Vygotsky (1978) and extending to modern cultural psychology (Tomasello, 2011).

The proposal is thus that our two evolutionary steps in the natural history of human morality reflect two fundamental and distinct forms of social engagement: the second-personal and the group-minded. And these two distinct forms of social engagement logically had to emerge in the order that they did. Although humans have always lived in social groups, it is almost inconceivable that modern humans' group-mindedness, in which the cultural group is understood as a collaborative enterprise that excludes free riders and competitors, could have emerged before an initial step of collaborative dyads or something similar. Each of these new forms of social engagement emerged from a generally similar evolutionary sequence: (1) changes in ecology (first the disappearance of individually obtainable foods and then increasing population sizes and group competition) led to (2) increases in interdependence and cooperation (first obligate collaborative foraging and then cultural organization for group survival), and then (3) coordination of these new forms of cooperation required new cognitive skills of shared intentionality (first joint intentionality and then collective intentionality), new social-interactive skills of cooperative competence (first second-personal competence and then cultural competence), and new processes of social self-regulation (first joint commitments and then moral self-governance). Each mode of social engagement thus represents a distinct set of biological adaptations for coping with a distinct form of social life.

The full story, as we have told it, comprises many more detailed twists and turns, of course. We have attempted to be as comprehensive as possible in detailing the many different aspects of human morality as they relate to the many different aspects of uniquely human collaboration and culture across evolutionary time. Because there are so few other evolutionary accounts with this expansive focus, we have so far made scant reference to other large-scale theories. But there are a number of other relevant accounts of the evolution of human cooperation and morality in general, and a broad survey of these will help to better situate the interdependence hypothesis within the current theoretical landscape.

Theories of the Evolution of Morality

Contemporary theories of the evolution of human morality fall into one of three very broad categories: evolutionary ethics, moral psychology, and gene–culture coevolution. We consider these each in turn.

The set of approaches grouped under the general rubric of evolutionary ethics focus on theoretical principles of cooperation in evolution and how they might apply to the human case. The foundational work from this perspective is Alexander's *The Biology of Moral Systems* (1987), which emphasizes processes of reciprocity, and especially, in the human case, indirect reciprocity. The evolutionary psychologists Cosmides and Tooby (2004) also focus on reciprocity and social exchange, with an emphasis on the special preparedness of human beings for detecting those who cheat in such exchanges. Sober and Wilson (1998) and de Waal (1996) also agree on the importance of reciprocity, but they emphasize empathy and sympathy as foundational to human cooperation and morality as well. De Waal (2006) finds both empathy and reciprocity in the behavior of nonhuman primates and believes that human morality emanates mainly from the greater cognitive and linguistic capacities of humans (though these are not further specified). Sober and Wilson (1998) believe that, in addition, humans underwent a special process of group selection that made it advantageous for individuals to be sympathetic and helpful to others in their group because of the advantages that then accrued to their group relative to other groups. Kitcher (2011) also emphasizes the role of altruism in human evolution but believes that it could not support genuinely human morality unless something else like "normative guidance" also evolved to acculturate developing individuals into the normative standards of the group.

Two more recent theories from this general perspective have emphasized the role of partner choice and social selection. Boehm (2012) focuses on the transition from dominance-based societies—that is, great apes in general and perhaps the earliest humans—to more egalitarian societies (see also Boehm, 1999). The major mechanism proposed is "selection by reputation" enforced by group punishment (which lowers the cost of punishment for each punisher) in the form of coalitions against cheaters and bullies of all kinds. The power of reputation is magnified with the advent of language, making possible gossip about the reputations of people one has never even met. Boehm speculates that the internalization of this process—the individual evaluating and possibly punishing herself through guilt—amounts to what is called a moral conscience (see also Sterelny, 2012). Baumard et al. (2013) are not so sanguine about the power of punishment (partner control) to create moral individuals. They focus on partner choice (as here, for mutualistic enterprises) and the way that it serves to create a predisposition in individuals for fairness based on individuals' relative contributions in procuring the spoils. In a biological market in which some individuals have more "leverage" than others—for example, by having special talents in some mutualistic enterprise—others will accede to that leverage only if they have no other outside options; if all have options, those who try to exploit their leverage will be socially selected against. Baumard et al. (2013, p. 65) attribute special importance to reputation and gossip in keeping individuals in line: a genuine morality evolved, they argue, because "the most cost-effective way of securing a good moral reputation may well consist in being a genuinely moral person," especially since everyone is on the lookout for individuals trying to fake it—though nowhere do they specify what "being a genuinely moral person" means.

From our perspective, all of these views have merit and have likely captured crucial aspects of the evolution of human morality. Almost all theorists agree that sympathy is an important part of the picture. But then the tendency is to try to cover everything else with one or another form of reciprocity. Our own view, as elaborated in Chapter 2, is that reciprocity is limited in explanatory power and that the notion of interdependence—which can also be thought of as various kinds of symbiosis—is much more powerful. But beyond that, the notion of reciprocity is not equipped to cover all the myriad aspects of human moral psychology, including such things as the making of joint or collective commitments and promises, creating and enforcing social norms, feeling resentment, and, most of all, self-regulating one's actions by feelings of responsibility, obligation, and guilt. These phenomena cannot simply be swept under the reciprocity rug. These social psychological phenomena would seem to derive from (1) a sense that I depend on others, just as they depend on me (interdependence), which paves the way evolutionarily for (2) a sense of "we-ness" (shared intentionality, especially in the form of a joint commitment) and self–other equivalence (impartiality) in all that we do. Our view is thus that the main limitation of these various accounts in evolutionary ethics is that they do not appreciate sufficiently the way that human morality depends on the senses of "we" and self–other equivalence as individuals interact socially with cooperative motives and attitudes on the proximate psychological level.

A second set of approaches comes from moral psychology. As the name implies, they focus less on evolutionary processes and more on proximate psychological mechanisms. Social psychology has a long tradition of investigating prosocial behavior (e.g., Darley and Latane, 1968; Batson, 1991), and developmental psychology has a long tradition of investigating the development of moral judgment in children (e.g., Piaget, 1932/1997; Kohlberg, 1981; Turiel, 1983). Neither of these traditions has had much to say about evolution, however. What has come to be known as moral psychology began with the neuropsychological investigations of Greene et al. (2001; see Greene, 2013, for a review) and has been tied more directly to biology and evolution. The focus in moral psychology has been mainly on people's judgments of harm, and the main methodology has been to ask them explicit questions, for example, about various permutations in the famous trolley car problem in which hypothetical agents have various kinds and amounts of control over when and how runaway trolley cars cause various kinds and amounts of damage. People across the globe answer these questions mostly similarly, perhaps suggesting a universal set of human intuitions about harm. Mikhail (2007) summarizes these intuitions from a legal point of view into two basic principles: a judgment of blame when someone intentionally harms someone else, but absolution when the intentional act that causes harm was aimed at something good (and there were no viable alternatives available). Researchers in this tradition, along with philosophers such as Nichols (2004) and Prinz (2007), have emphasized the role of emotions and intuitions—rather than moral reasoning—in humans' moral decision making.

By far the most thorough and detailed account in the moral psychology tradition is that of Haidt (2012). Like other moral psychologists, Haidt focuses on purported innate predispositions that lead to quick, intuitive moral judgments, often laden with emotion. He has also proposed that the kind of explicit reasoning about moral issues in which human beings regularly engage is actually post hoc and justificatory only; that is to say, it rationalizes and justifies the already made intuitive judgment. Its function is to persuade others that this judgment is the best one and so they should support it in any kind of dispute. In addition, Haidt goes beyond the singular issue of harm and tackles explicitly the diversity of moral judgments that people make, including people from diverse cultural backgrounds. Whereas other moral psychologists have noted non-Western moral intuitions and cultural variation, Haidt has created a theory based on them. He proposes that human moral judgment universally rests on five pillars: care/harm, fairness/cheating, loyalty/betrayal, authority/subversion, and sanctity/degradation. Also using mainly explicit questioning, Haidt has shown experimentally that the diversity of people's moral judgments can mostly be accounted for by various weightings of these different moral pillars in particular situations. From the standpoint of evolution, Haidt relies strongly on processes of group or multilevel selection, in which the adaptiveness of individuals is intimately bound to how their group does in competition with other groups, with individuals who show prosocial or moral tendencies creating more effective social groups.

Rand et al. (2012; see also Greene, 2013) have categorized all of this research, as well as their research on prosocial behavior, with the well-known distinction between type I and type II cognitive processes. Type I processes are quick, intuitive judgments, often based on emotion, whereas type II processes are slow and deliberate judgments arrived at via explicit reasoning. Clearly, most subjects in most moral psychology experiments are relying on type I processes. From the point of view of evolution, quick, intuitive judgments backed by strong emotions are quite likely adaptations for dealing with biologically urgent situations in which there is no time to think and/or in which too much thinking might just confuse the matter. Much of the research in moral psychology is thus from a nativist perspective in which the goal is to show that the kinds of conscious thinking that humans believe they are using to make a decision are actually not determinative; rather, what is determinative are innate predispositions of which we are mostly unaware.

Few moral psychologists have speculated about the adaptive conditions that might have led to humans' unique moral psychology, the main exceptions being Haidt (2012) and Greene (2013), who both focus on the important role of human cultural groups and individuals' resulting group-mindedness. And, crucially, few moral psychologists have paid much attention to other aspects of morality beyond emotion-based intuitive judgments, in particular, those involving rational decision making and the resulting sense of obligation. Our view, then, is that although the field of moral psychology has been revolutionary in identifying and investigating the type I decision-making processes that account for a good proportion of humans' actual moral judgments, it has not attended sufficiently to those aspects of human morality that emanate from the unique ways that humans understand their social worlds rationally and make rational decisions based on this understanding.

The third set of approaches to the evolution of human morality focuses more specifically on the important role of culture. Although it is not clear that anyone espouses a purely cultural view in which biological evolution plays no role (although see Prinz, 2012), nevertheless many cultural anthropologists and cultural theorists choose to emphasize the way that individuals are enculturated into a particular kind of "moral code," and these codes may differ quite radically across cultures. From a more moral psychological point of view, for example, Shweder et al. (1987; see also Miller, 1994) criticize Western views of moral psychology in which autonomous individuals freely agree and assent to their moral code. More accurate, they argue, is a view in which culture is the major force in creating autonomous individuals in the first place before any agreements are possible. Thus, in the Hindu cultures of India that Shweder et al. (1987) have studied most intensively, individuals do not think of themselves as totally autonomous agents; rather, they see themselves as operating in subordination to natural law and objective obligation as set forth in the various practices and doctrines of their culture. The overall conclusion of Shweder et al. (1987, p. 35) is that "there is relatively little evidence for a spontaneous universal childhood morality not related to adult attitudes and doctrines."

More congenial to the current view are approaches that explore the role of culture in its evolutionary context. Richerson and Boyd (2005), for example, emphasize the critical role of cultural group selection in the evolution of human cooperation and morality. The idea, as noted above, is that once cultures began evolving, different cultural groups could compete with one another, such that those with the most cooperative individuals would very likely do best. There were thus selection pressures within each cultural group for imitation and conformity, especially of successful individuals, and immigrants from other groups imitated as well (which makes cultural group selection more plausible than biological group selection, in which immigration undermines the genetic integrity of the group). Humans thus have evolved, in a process of gene–culture coevolution, both cooperative tendencies and "tribal instincts" for living and functioning effectively in cultural groups. Bowles and Gintis (2012) emphasize that the individual psychology involved is a tendency toward "strong reciprocity" in which individuals both cooperate with others and punish noncooperators.

In the current account we have invoked processes of cultural group selection to explain the ratcheting up humans' special sense of group-mindedness, at the end of the second step of our story. But it is clear that cultural group selection can explain only why the particular social norms and institutions of particular cultural groups have prevailed in the last few tens of thousands of years of human evolution; it cannot explain the species-universal skills and motivations for creating social norms and institutions in the first place. These species-universal skills and motivations would have needed to be in place before such processes of cultural evolution could have gotten started, and we have proposed one account of how this might have happened in terms of the social and moral psychology involved. Cultural group selection is thus clearly an important part of the overall story, but only, we would claim, at the very end.

Compared with these various approaches in evolutionary ethics, moral psychology, and gene–culture coevolution, the current account has attempted to be more comprehensive. This manifests in three key ways. The first is that it hypothesizes two distinct steps that were necessary for the evolution of human morality: a first step involving dyadic interactions in small-group contexts, and a second step involving group-minded interactions in cultural contexts. The second is that it hypothesizes with some specificity the changes in human socioecology that provided the adaptive context for each of these two stages: first, the emergence of obligate collaborative foraging with partner choice, and second, the emergence of tribally organized cultural groups that competed with other similar groups for resources. The third is that it characterizes human moral psychology as constituting the cooperatively rational way for individuals to treat one another in these new social contexts: at the first step individuals with skills of joint intentionality and second-personal agency made joint commitments based on a kind of cooperative rationality, and at the second step individuals with skills of collective intentionality and cultural agency made collective commitments to the norms and institutions of their social group based on a kind of cultural rationality. All of this has relied heavily on concepts from moral philosophy and social theory, and this is perhaps a distinguishing feature of our evolutionary account as well.

Shared Intentionality and Morality

We have attempted in this account to ground human morality in human cooperation without reducing it to it. We have attempted this empirically by documenting the many different ways in which human cooperation differs from the cooperation of other primate species. We have attempted this theoretically by connecting the species-unique characteristics of human cooperation to genuine moral decisions not aimed exclusively at strategic ends. Consistent with this naturalistic approach, we have made these connections via processes of human psychology. To enable a more detailed comparison with the many other views just enumerated, then, let us recapitulate in broad outline some of the most important points in this speculative evolutionary narrative:

- Great apes are instrumentally rational beings that make more or less intelligent decisions. Their social life is structured mainly by competition, with most of their cooperation (e.g., in coalitions and alliances) serving competitive ends. The strongest social selection is thus for good competitors. They show sympathy for close relatives and "friends," such as coalition partners (e.g., preferentially grooming them), and sometimes for other friends in need if the costs of helping are not too great.
- Chimpanzees and bonobos hunt in groups for small mammals, but there are few signs of a cooperative structuring of this activity. They do not share the spoils freely among participants (only in response to harassment or, sometimes, among coalitional partners). Nor do they make any efforts to exclude free riders from the spoils. There seems to be no sense of

fairness. In general, great apes do not seem to be specifically adapted—beyond other mammals—for "cooperative ventures for mutual gain."

- Early humans (by at least four hundred thousand years before present) were forced into an ecological niche of obligate collaborative foraging, which made individuals strongly interdependent for survival, and thus sympathetically concerned for the well-being of potential partners. They cognitively recognized this interdependence, so it was an integral part of their flexible and strategic—and now cooperative—rationality.
- Early humans' collaborative activities thus came to have the dual-level structuring of joint intentionality: a joint agent "we" comprising two interdependent partners, "I" and "you" (reciprocally defined). Each partner had his role in the collaboration, and both partners knew together in common ground the ideal way those roles must be played for joint success. Because partners understood that both of them were required for joint success, and because they knew that their role ideals were agent independent and interchangeable, there arose a sense of partner (self–other) equivalence from a "bird's eye view."
- Early humans' collaborative activities took place in the context of partner choice in which potential partners evaluated others for their cooperativeness. Unlike great apes, early humans knew that they were being evaluated by others as well—and indeed, they could reverse roles and simulate others' evaluations—so they knew their value to others as partners. In combination with the sense of partner equivalence, this led to a sense of mutual respect between partners. In excluding free riders, partners also—again, in combination with the sense of partner equivalence—evolved a sense of the equal deservingness of partners, not free riders, to an equal sharing of the spoils. Treating others as equally deserving partners turned early human individuals into second-personal agents with cooperative identities.
- Early humans could use their joint agent "we" to make a joint commitment to self-regulate the collaborative activity. This joint commitment was created via the second-personal address of cooperative communication and assured that both partners would persist through distractions and temptations until both received their just deserts. Deviations from ideal role performance were met with second-personal protest, which respectfully requested that the deviator self-correct, which she then had to do to maintain her cooperative identity as a virtuous partner. She self-

corrected not just out of a fear of punishment but because the protest was legitimate (deserved) since it came from "us" in our cooperative identities. An internalization of this process led each of the partners to feel a sense of second-personal responsibility to the other and second-personal guilt when they did not live up to this responsibility. This whole "we > me" way of self-regulating constituted a radically new form of cooperative rationality, in the sense that each partner freely relinquished some personal control over his actions to the joint agent of which he was a constitutive part. The result: two second-personal agents self-regulating their collaboration via mutually agreed-upon and impartial normative ideals.

- At some point, modern humans began living in larger, more coherent, and tribally structured cultural groups that competed with other such groups for resources (by at least one hundred thousand years before present). This led to a distinct group-mindedness in which individuals knew that they were dependent on the group more than the group was dependent on them, so they conformed to its strictures. The interdependence of in-group members led them to be especially sympathetic and loyal to one another but unhelpful to and mistrustful of all outgroup barbarians.
- All of modern humans' cultural interactions were structured by collective intentionality, underlain by a sense of cultural common ground. The result was conventional cultural practices—everything from making a spear to raising children—which, in theory, were effected in the same ideal way by any culturally competent individual. Roles in these practices—analogous to the individual roles in a dyadic collaboration—were thus fully agent independent, and the role of simply being a contributing group member led to a sense of distributive justice that was, at least in theory, fully impartial within the group. This new way of operating created the conditions for individuals to construct a fully agent-independent, objective, and impartial view of the world, which resulted in, among many other things, an objectification of role ideals into the right and wrong ways of doing things.
- Modern humans' processes of social control—as transformations of early humans' second-personal protest—manifested in their social norms and institutions. Individuals were born into this preexisting cultural reality that was presented to them most often in the voice of a generic normative language, such as "one must do it like this" or "this is the right way

to do it," which seemed to come not from the speaker as an individual but from an objective world of objective facts and values. This objective world of right and wrong was handed down to individuals by their venerated ancestors in the culture. To live up to the positive aspirations embodied in these values was to be virtuous, and to not live up to them was to undermine the cooperative rationality of our cultural life, and so one's own cultural and moral identity.

- Modern human individuals did not just create joint commitments with partners to self-regulate the dyad; they also bought into the social contracts that already existed in their culture (i.e., its norms and institutions) and used them to self-regulate. Again, they did this not just out of a fear of punishment but out of a respect for the legitimacy of these supraindividual social structures with which they identified themselves as, in effect, coauthors. The sense of obligation to act in concert with one's cultural group thus represented still another, scaled-up form of cooperative rationality: the cultural rationality of not just joint agency but cultural agency. The self-regulation of "we > me" in modern humans therefore took the form of moral self-governance: the individual respecting and internalizing the objective values of the group—while at the same time questioning these and, where appropriate, providing her "reflective endorsement"—as part of her moral identity. The individual sought to maintain her moral identity in the face of discrepant behavior by creative interpretations and justifications of that behavior grounded in the shared values of the cultural group, or by displaying guilt in ways that identified with the group's moral judgments.
- At some point, different cultural groups began creating somewhat different conventions, norms, and institutions, and groups with more effective versions outcompeted other groups (cultural group selection, with gene–culture coevolution selecting for cooperative individuals), with the process intensifying in civil societies with codified laws and organized religion.

The interdependence hypothesis for the evolution of human morality is thus about the proximate psychological mechanisms that emerged in support of early and modern humans' newly interdependent and cooperative ways of life, which of course had as their evolutionary starting point great apes' already very social ways of life. A highly schematic graphic summary of the overall account is depicted in Figure 5.1. The arrows are intended to reflect genuine qualitative transformations in which one way of doing things is transformed, via processes of natural and social selection—and perhaps in some cases cultural creation and learning—into a different way of doing them.

But beyond this description, the more ambitious goal of the current account is to provide an explanation, an evolutionary explanation, for how the human species transformed great ape strategic cooperation into genuine human morality. Since evolutionary explanations presuppose that individuals generally act to enhance their own reproductive fitness, the idea of natural selection producing genuinely moral individuals—who care about the welfare of others, who see others' concerns as equal to their own, and who relinquish control of their individual actions to a self-regulating plural agent "we"—might seem like pulling a rabbit out of a hat (as one reviewer put it). But evolution pulls rabbits out of hats all the time—in the sense that it produces evolutionary novelties all the time—so the challenge is to specify how it did it in this case.

The most fundamental theoretical concept is, of course, the concept of interdependence. We have argued that, whereas reciprocity (direct and/or indirect) may be an appropriate description of certain behavioral patterns in terms of a cost-benefit analysis, it does not help much in explaining human moral psychology. Rather than focusing on social interactions in terms of the objective costs and benefits to the actors involved, it is much more helpful, we would argue, to focus on the dependencies among the actors involved and, at least for current purposes, how the actors understand those dependencies. Dependency is the normal way that evolutionary biologists conceptualize interactions between different species in terms of various kinds of symbioses (e.g., parasites and commensualists depend asymmetrically on their hosts, whereas symbionts depend on one another symmetrically), and the notion of interdependence simply adopts this way of conceptualizing things for conspecific individuals within a social group. This dependency conceptualization avoids the many well-known problems of reciprocal altruism (especially the undermining effect of cheating), and, we would argue, it also provides a congenial framework for explaining the evolutionary origins of human moral psychology.

It does this in two main ways. First, the dependency conceptualization makes individuals caring about the welfare of others, and so helping them, a natural part of social life. In this way of looking at things, the psychological process is not one in which the cooperating individual is reciprocating, in the

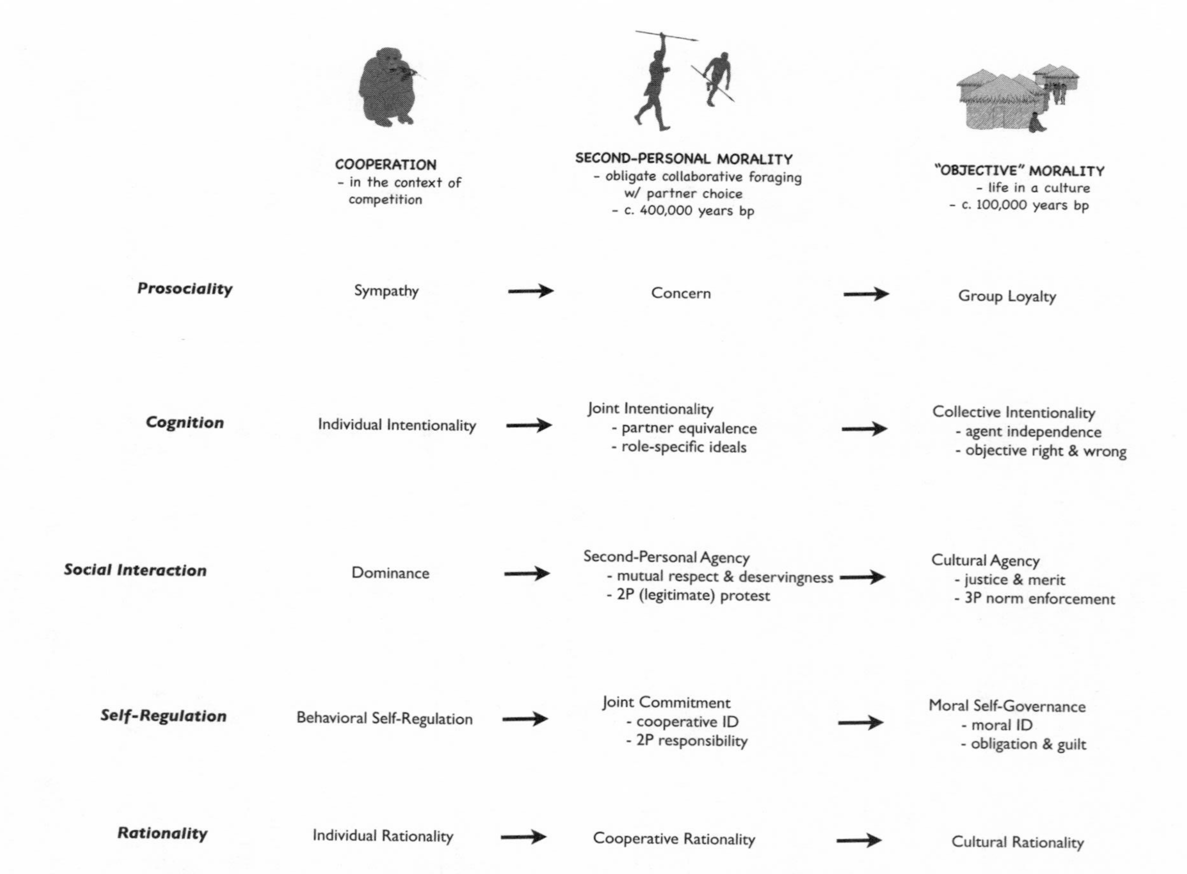

FIGURE 5.1. Summary of the interdependence hypothesis for the evolution of human morality. 2P, second-personal; 3P, third-party; ID, identity.

sense that he is repaying another individual for a past act; rather, he is investing, with an eye to the future, in the well-being of an individual on whom he depends. Obviously, as explicated in Chapter 2, there is a mathematics involved here such that individuals cannot overdo the self-sacrifice—just as they cannot overdo the self-sacrifice in helping kin—and of course, the fact is that the individual does end up benefiting in the long run. Some theorists might therefore conclude that we should not then think of helping behavior underlain by interdependence as caring about the welfare of others at all. But this is to confuse the evolutionary and psychological levels of analysis; at the psychological level the caring is in many cases totally genuine.

Let us then attempt to keep the evolutionary and psychological levels clearly distinct. Focusing on the human case, there are many situations in which individuals understand the logic of interdependence reasonably well and so help others strategically (and this would, of course, be adaptive on the evolutionary level as well). For example, an individual might help his collaborative partner because he knows that this furthers progress toward their joint goal, or she might help an individual whom she knows to be important to her fitness (e.g., a mate) or with whom it is important for her to maintain a good reputation. Juxtaposing the psychological and evolutionary levels terminologically, then, we may call such behaviors *strategic-adaptive,* as they are self-serving, and so not moral, on both levels. But we have argued and presented evidence that in many instances even young children help others based on a genuine concern for their welfare, without any strategic calculations on the psychological level at all. These are the cases that most people would consider to embody something like a morality of sympathy. If indeed the evolutionary origin of such caring and helping is interdependence, then these behaviors do in fact have an evolutionary payoff to the individual, though he may not know it. Juxtaposing the psychological and evolutionary terminology in this case, we may then call such actions *moral-adaptive,* in the sense that the individual is helping based on a genuine feeling of sympathetic concern, even though, unbeknownst to him, there is an adaptive payoff. The first way that interdependence helps to explain the evolutionary origins of human moral psychology, therefore, may be considered a version of what some people have called a "mistake," in the sense that the individual believes himself to be acting out of a genuine concern for the other even though, on the evolutionary level, he himself benefits. In any case, what we are envisioning here are acts in which the psychology is genuinely moral irrespective of any evolutionary payoffs.

The second way that interdependence helps to explain the evolutionary origins of human moral psychology is more indirect, and it is not based on a mistake but, rather, on a rational assessment of the way things are. The initial step was that early humans developed a new set of cognitive skills for coordinating their collaborative and then, later, their cultural activities: shared intentionality. Because of the dual-level structure of shared intentional activities—"we" at the top level is in control of "you" and "me" (perspectivally defined) at the bottom level—participants developed a common-ground understanding of how each role in a shared intentional activity had to be played for shared success, no matter who played that role. Participation in shared intentional activities also entailed participants taking one another's perspective and, indeed, trying to manipulate one another's perspective through acts of cooperative communication. As an integral part of operating with person-independent role ideals, then, individuals developed a sense of participant, or self–other, equivalence in the collaborative or cultural activity from a "bird's eye view." The role reversal entailed in this conceptualization of a collaborative or cultural activity meant that participants could—perhaps even could not help but—monitor and evaluate their own role performance in the same way that they monitored and evaluated others'.

The point is that cognitive skills of shared intentionality (including cooperative communication) are evolved adaptations enabling individuals to better coordinate their collaborative and cultural activities (Tomasello, 2014). They do not concern morality directly; rather, they structure the way that individuals understand their collaborative interactions and the participants in them. Our argument is that as early human individuals interacted with others in the context of partner choice, they were at some point already equipped with a new understanding of self–other equivalence, and this led to something radically new in the natural world. As they interacted in a relatively egalitarian biological market, it was in the interest of individuals neither to tolerate being taken advantage of nor to take inordinate advantage of others. By itself, this would simply generate a balance of power. But in combination with an already developed conceptualization of self–other equivalence—collaborative partners (but not free riders) as equally important to instrumental success and equally evaluable by the same ideal standards—the respect became not about power but about something like "deservingness": if we are equivalent in the collaborative process, then we *deserve* equal treatment and benefits.

And so the first leg of the rabbit has been pulled out of the hat. A situation that other primates would perceive as constituted by power, early human individuals perceived as constituted by some kind of status—indeed, equal status—in terms of respect and deservingness (Darwall's [1997] recognition respect). What was responsible for this conceptual transformation was an understanding of self–other partner equivalence, evolved as a purely cognitive judgment in the context of attempts to coordinate collaborative activities. Coming to perceive and treat others with respect as equally deserving second-personal agents was thus a joint result of both special social-interactive conditions—especially partner choice with certain parameters—and special cognitive capacities previously evolved for other functions. If the fundamental moral concept here is "deservingness"—with an individual feeling resentment for not being treated with the respect she deserves—then it was enabled by humans' earliest inklings of an impartial perspective that, to repeat, was evolved for purely coordinative purposes. Continuing our labeling convention indicating simultaneously the evolutionary and psychological levels, we may then label the creation of equally deserving second-personal agents as *moral-structural*. The term *structural* indicates that, on the evolutionary level, this dimension of human morality was not originally selected to serve in this function; it came into existence in the service of other functions. Of course, this would not work if a recognition of self–other equivalence was maladaptive in the social-interactive contexts we are considering, but it could easily be a cost-neutral "spandrel" that cognitively structured such contexts.

We may also characterize the evolutionary origins of humans' senses of commitment, responsibility, obligation, and legitimacy as also of the moral-structural variety. The social-interactive context for making a joint or collective commitment is to reduce the risk of a collaborative enterprise, and so the partners "agree" that their coconstructed plural agent will self-regulate their joint or collective agency ("we > me"). Agreeing to this kind of self-regulation, of course, has a strategic dimension—to not agree would be to risk one's status as an attractive cooperative partner—but at the same time it has a moral dimension as well. The moral dimension is that participants view self-regulation by the plural agent, including any sanctions it may mete out, as legitimate (deserved), so it is part of the participants' cooperative or moral identities. This sense of legitimacy is thus based both on the impartiality of the way it works—we agreed to sanction whichever of us does not live up to our role ideals (under a kind of "veil of ignorance")—and on the concomitant possibility of role reversal evaluations such that individuals cannot help but evaluate themselves impartially, in the same way that they evaluate others. Both this impartiality of the joint commitment and this role reversal evaluation are, ultimately, based once more in a sense of self–other equivalence. This way of thinking and operating thus constitutes a new kind of cooperative rationality based not only on instrumental success but also on a sense of fairness in one's dealings with others, and it comes—because it is still a form of rationality and decision making—with a normative sense of obligation to one's partner or compatriot to live up to our shared ideals. Thus, once again, we have a social-interactive arena that is structured by a preexisting way of conceptualizing both self and others from a kind of impartial "birds eye view" perspective, so we may refer to all of this again as moral-structural.

And finally, we may also characterize most of the elements in the move of modern humans from a natural, second-personal morality to a cultural, "objective" morality as moral-structural. Most important, the objectification of the values underlying social norms into the values of right and wrong—the most distinctive feature of modern human moral psychology—was the result of processes of collective intentionality, in which the self–other equivalence and role reversal evaluation characteristic of early humans were scaled up into fully agent-independent thinking and impartial moral judgments as viewed from a perspectiveless "nowhere." Modern human collective intentionality comprised cognitive adaptations facilitating individuals' ability to function effectively in a group-minded world of conventional cultural practices, norms, and institutions, and so it was not in itself any kind of moral motivation or judgment. And without cognitive skills of collective intentionality, there could be no processes of moral self-regulation such as the reflective endorsement of moral decisions and a full-fledged sense of guilt. We may therefore say once again that the objectification of the instrumental values underlying cultural practices and social norms into moral judgments of right and wrong was moral-structural.

An important implication of this way of looking at things is that morality is not a domain of activity with an isolated evolutionary history—it is not a module (whatever that means); rather, it is a complex result of many different processes, each with its own evolutionary history. Human morality is the way that humans have come to interact with one another in the context of certain cognitive insights about how the world, including plural agency, works.

Treating others as equally deserving as oneself in dividing resources fairly, or chastising oneself in the same way one would chastise others for violating a social norm, reflects a genuine morality emanating from the individual's perception of himself as equivalent to others in relevant respects, that is, from an impartial point of view. In this way the current account circumvents the specter of natural selection favoring individuals who are motivated to subordinate or equate their own interests to those of others evolutionarily. Instead, individuals are acting with a kind of cooperative rationality based on an accurate recognition of social reality, and this is at least viable on the evolutionary level because of the interdependent sociality of human individuals. We may summarize the overall theoretical structure of the interdependence hypothesis as in Figure 5.2.

And so now the rabbit has been pulled fully from the hat, and the sense of magic is palpable. We have not explained any of the components in this account fully, and there is a fair amount of hand waving when we get down to the evolutionary details—and in some cases certain conditions (e.g., a biological market of a certain type) are basically assumed to make everything work.

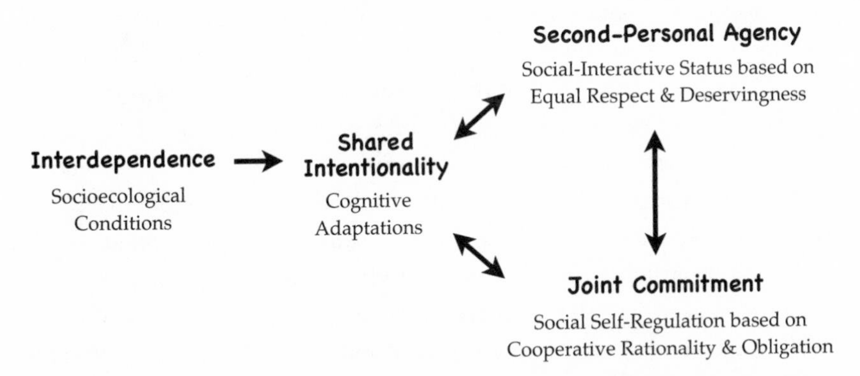

FIGURE 5.2. Overall theoretical structure of the interdependence hypothesis. The attempt is to use terminology neutral between the two evolutionary steps, but in many cases no overarching term has been introduced for this function. Thus, for present purposes, *second-personal agency* includes cultural agency, *joint commitment* includes collective commitment, *cooperative rationality* includes cultural rationality, and *obligation* includes second-personal responsibility. *Interdependence* and *shared intentionality* are not moral phenomena, but when certain kinds of cooperative interactions take place in the wake of their evolution, the result is individuals who view one another with equal respect and deservingness and who feel an obligation to live up to the social commitments they have made (or affirmed) with one another.

But given that we are attempting here an imaginative reconstruction of historical events many thousands of millennia in the past—with little in the way of artifacts or other paleoanthropological data to help—it is going to have to do for now. What we have attempted in this final discussion is simply to provide a very general explanatory account to show that genuinely moral beings, who are genuinely concerned about the well-being of others and who genuinely feel that the interests of others are in some sense equal to their own, could plausibly have emerged during the course of human natural history without violating any of the basic principles of evolution by means of natural selection.

The Role of Ontogeny

At both of our evolutionary steps, human children became moral beings only gradually and within a social context. But biological adaptations may express themselves in ontogeny in myriad ways, involving the social context in myriad ways (including no way at all), so we are still left with questions about the ontogenetic process.

As just one approach to these questions, we may consider the moral development of contemporary human children, about whom there is a wealth of empirical research (a selected set of which is cited here). First of all, despite the evolutionary perspective of the current approach, there is absolutely no question that much of contemporary children's moral behavior is culturally learned from adults and that many of their moral attitudes and judgments are internalized from interactions in which adults articulate and enforce social norms. As evidence, there are many well-documented cultural differences in the moral behavior and judgments of children growing up in different cultures, some of which have been reviewed here, and the research of Shweder et al. (1987), Haidt (2012), and others has demonstrated just how different are the moral sensibilities of people coming to maturity in different cultural and religious contexts.

But the household pets growing up in these same cultural and religious contexts do not thereby become moral beings. And so, second of all, there is also absolutely no question that human children are biologically prepared for the process as well. In his analysis of the evolution of human morality, Joyce (2006, p. 137) emphasizes this fact by problematizing the potential role of the social environment: "It is puzzling what there even *could be* in the environment—even a rich and varied environment—for a generalized learning mechanism to latch on to in order to develop the idea of a moral transgression. . . . Administer as many punitive beatings as you like to a general-purpose intelligence; at what point does this alone teach it the idea of a *moral transgression?*" As always in developmental analyses, the question is how biological adaptations express themselves in particular environmental contexts over ontogenetic time.

Our attempt here has been mainly to specify key aspects of humans' biological preparedness for becoming, during an extended ontogenetic process, moral beings. As applied to ontogeny specifically, the proposal of Tomasello and Vaish (2013) is that, despite some important differences, contemporary children actually pass through two developmental stages not unlike those we have posited here for phylogeny. Chapter 3 cited a number of studies showing that young children from one to three years of age behave morally toward other individuals in a variety of circumstances. They help them spontaneously, and this helpfulness is intrinsically motivated (Hepach et al., 2012). They share with others, and when they are collaborating with others they share with their partners equally, even if this means giving up a resource themselves (Hamann et al., 2011). They show some signs of honoring and expecting others to honor joint commitments, even to the point of openly acknowledging when they break one (Hamann et al., 2012; Gräfenhain et al., 2009). One- to three-year-old children thus behave toward others, at least sometimes, with sympathy and a sense of fairness. They possess a natural, second-personal morality that emerges as they interact with others in the first three years of life.

Of course, contemporary one- to three-year-old children differ significantly from the early humans in the first step of our evolutionary story. They are born into a cultural world, and they have modern human skills and motivations for imitation and cultural learning, including conforming to the actions and imperatives of others. Nevertheless, our claim is that before three years of age young children do not yet understand social norms as the shared expectations of "our" social group. That is, despite their tendency to follow the imperatives of others (which those others may understand as social norms), it is only after three years of age that young children understand social norms per se. As evidence is the fact that it is only at this age that children, for the first time, begin actively to enforce social norms on others. It is this third-party norm enforcement that expresses the child's sense of cultural identity that "we" should all do things in the right way (and make sure that others do as well; Schmidt and Tomasello, 2012). Additional evidence is the fact that children understand themselves as members of a group per se—based on such things as physical and behavioral resemblance (and as opposed to simply differentiating familiar versus strange people)—and show loyalty to the group only after three years of age (Dunham et al., 2008). We would thus argue that young children start becoming part of a group-minded, cultural morality based on a cultural identity only sometime after three years of age.

It is therefore at this second developmental step, beginning at three years of age, that specific types of social and cultural interactions and instruction from adults in the culture become critical. First of all, of course, children must learn from adults the specific conventional and moral norms of their culture; that is clear. In addition, however, there is a rich history of developmental research into the socialization practices that encourage children to internalize these norms and so to begin making cooperative and moral decisions based on these norms for themselves (see Turiel, 2006, for a comprehensive review). A key finding is that more authoritarian parenting styles lead to less internalization of values and so to more strategic norm following, whereas more inductive parenting styles, focused on providing children with reasons for actions, lead to more internalization of values and so to moral self-regulation (Hoffman, 2000).

But following Piaget (1932/1997), perhaps even more important is what children learn through their interactions with peers. Consistent with the current analysis in which following social norms is all and only about conformity, Piaget contends that conforming to norms based only on adult authority and fear of punishment is merely prudential. To develop a genuine morality, something else is needed. Piaget argues that it is through their interactions with peers—individuals with equal power with whom it is necessary to argue and negotiate on an equal footing—that children develop genuinely moral concepts and attitudes. In current terms, children can learn to relate to others as equally deserving second-personal agents only from interacting with others who are in fact equal. And so children can learn through their interactions with peers, but not from adult instruction, how to resolve moral conflicts by creatively applying their naturally developing moral attitudes of sympathy and fairness. And they can learn through their interactions with peers, but not from adult instruction, how to resolve conflicts between different social norms and so to create for themselves a personal moral identity. In Piaget's view, and in our view, obeying authority is one thing, but being concerned with others and learning to treat them with respect and fairness is another, and it can only come from interactions with peers. Morality is how we work things out with others by means *other* than power and authority, and, while we may get some hints from previous generations, in the end it is we of this current generation who must negotiate our own moral relations.

In general, the prediction from the current account would be that before three years of age young children across all human cultures are very similar in terms of both cooperation and morality, and only after three years of age will they begin to identify with the particular social norms of their cultural group and so to construct a cultural morality. The problem is that there are very few cross-cultural studies of children's social behavior before three years of age. The one notable exception is the study of Callaghan et al. (2011), who performed a series of experiments with children from one Western, industrialized culture and two small-scale, traditional cultures (in India and Peru). In tasks of helping and collaboration, they found similar behaviors at similar ages across all three cultures. This pattern is at least consistent with current predictions. Further support for this hypothesized developmental trajectory is provided by cross-cultural research on distributive justice in the sharing of resources (see review and citations in Chapter 4), in which children from different cultures begin as more similar and then gradually track their culture's norms over development. As a corollary, we also hypothesize increasing cross-cultural differences over development based not just on the different content of different cultural norms but also on the way that different parenting styles or peer interactions encourage the internalization of moral values to differing degrees.

Human beings thus develop over their individual ontogenies a variety of skills, emotions, motivations, values, and attitudes—some biologically inherited, some culturally inherited, and some individually constructed—that influence their moral decision making in important ways. But still individuals are not determined by either their biology or their culture to act in any particular way on any particular occasion; indeed, in many complex situations there is no optimal solution that either biology or culture could possibly have anticipated. No, for better or for worse, there is no alternative to human individuals—however biologically and culturally equipped—making their own moral decisions.

Conclusion

> Ethical ideas, within any given human society, arise in the consciousness of the individual members of that society from the fact of the common social dependence of all of these individuals upon one another.
>
> —GEORGE HERBERT MEAD, *MIND, SELF, AND SOCIETY*

Quite often in the social sciences the human individual is portrayed as a rational maximizer, *homo economicus,* driven exclusively by the possibility of concrete personal gain. This psychological model is based explicitly on the supposed motivations and behavior of individuals acting within a capitalist market. But it is clear, in the broader sweep of human evolution and history—beginning with the egalitarian and communal hunter-gatherer societies that characterized the species for the first 95 percent of its existence—that capitalistic markets are cooperative cultural institutions. They are created by a set of cooperative conventions and norms in which individuals agree to follow a set of rules that, somewhat paradoxically in this case, empower them in certain contexts to pursue personal gain to the exclusion of everything else. The rules that empower individual self-interest in capitalist markets are thus like the rules that empower a tennis player's self-interest in defeating an opponent, that is, within the context of the cooperative rules that constitute the game in the first place. It is only if one neglects the cultural-institutional context of human behavior that one can hallucinate the competitive cart as leading the cooperative horse.

But certainly an account of human behavior grounded in evolution, such as the current account, must take individual self-interest as primary, well before and much more fundamental than cooperative social interaction? Well, yes and no. The logic of natural selection, of course, stipulates that organisms do things that increase, or at least do not decrease, their reproductive fitness.

One could call that self-interest. But what we normally refer to as self-interest is an individual making an active choice to favor itself over others. The vast majority of life forms on planet Earth are making no such choice. They are simply acting instrumentally toward their goals, and this means, for successful animals, that those goals are compatible with their continued survival and reproduction. But they have no psychological mechanisms specifying that they should favor themselves over others; the question simply does not arise. To say that they are acting out of self-interest is to confuse ultimate causation with proximate mechanism.

But for some socially complex animals, including primates and perhaps other mammals, the question of self-interest does arise. We have thus argued and presented evidence that great apes, on some limited occasions, will favor others over themselves. There may be an evolutionary explanation for this behavior in terms of some kind of payback, but the acting organism knows nothing of this payback; she is simply, for example, helping her friend by grooming her or joining her side in a fight, because she is her friend. But we have also argued and presented evidence that on other occasions great apes may favor themselves over others, for example, by hogging a resource even when they know that another individual wants it too. In the current focus on proximate psychological mechanisms, only something like this may be called acting out of self-interest. In general, because there is good experimental evidence that great apes often do things to benefit themselves over others (even when they know that they are thereby thwarting the other's goal pursuit), we may say that great apes are often, perhaps most often, acting out of self-interest.

Obviously, humans also have the capacity to act out of self-interest and quite often do. But we have argued and presented evidence that quite often, as well, even young children are genuinely concerned about the welfare of others without strategic calculation: they help others reach their goals, they share resources with them fairly, they make joint commitments and ask permission to break them, they act toward a "we" or group interest, they enforce social norms on third parties on the basis of presumably group-minded motives, and they have genuinely moral emotions—from sympathy to resentment to loyalty to guilt—that do not spring from any self-interested calculations at all. These empirical findings—and many others in other disciplines (see Bowles and Gintis, 2012)—suggest that human beings have evolved biologically to value others and to invest in their well-being. We have argued here that the explanation for this fact is that human individuals recognize their interdependence with others and the implications this has for their social decision making. They have become *cooperatively* rational in that they factor into their decision making (1) that helping partners and compatriots whenever possible is the right thing to do, (2) that others are equally as real and deserving as themselves (and this same recognition may be expected in return), and (3) that a "we" created by a social commitment makes legitimate decisions for the self and valued others, which creates legitimate obligations among persons with moral identities in moral communities.

From the individual's point of view, this is all very genuine: the moral judgments of both self and other in the moral community are, on the whole, legitimate and deserved. We would therefore speculate that most contemporary adult human beings, if given Plato's Ring of Gyges, which would make their actions invisible to others, would still behave morally most of the time. Invisible humans would undoubtedly break many social norms that have no connection to their second-personal morality. And they would undoubtedly behave immorally if their selfish motivations were strong enough. But in the absence of overwhelming selfish desires, invisible humans would most often help others and treat them fairly, and even feel guilty if they did not do so—assuming, of course, that they viewed them as a part of their moral community. And this would hold, we hypothesize, for all individuals in all moral communities across all cultures. What would differ is simply the ways in which people from different cultures, given the different social and institutional settings within which they live their lives, understand what are the right and wrong ways to do things in particular contexts, and who is considered a part of the moral community.

Our account is thus grounded in a natural second-personal morality. But in the contemporary world this natural morality is embedded in a cultural morality of social norms, and these have been crafted at different historical periods for different recurrent situations, so they sometimes conflict. In facing a novel situation, then, the individual must create his own moral principles to help adjudicate among these norms and so make decisions that enable him to preserve his moral identity. The problem is that there seem to exist genuine moral dilemmas that appear as a kind of Necker cube: moral in one way when looked at from one angle but moral in a different way, or even immoral, when looked at from a different angle. They have no general solution; they simply represent a collision of moral forces, and the individual must find some way to harmonize them, almost always by suppressing or overriding something (Nagel, 1986, 1991). What alternative is there, then, to an explanation of human morality in terms of a variegated history of biological adaptations and cultural creations that each work well in their respective "proper domains" but that collide with one another in novel situations that neither nature nor culture could foresee?

It is clear that many people will think that what we have painted here is an unrealistically rosy picture of human cooperation and morality. Where we see sympathy and a sense of equality, they would propose clever strategies for fulfilling selfish interests. When I donate money to a beggar on the street, they would argue, what I am *really* doing is attempting to enhance my reputation in the eyes of others. But why the *really?* Why can I not be doing both? Nothing makes for a better behavioral decision than something that achieves two goals at once: I help the poor person for whom I feel genuine concern, and I enhance my reputation at the same time—win-win. The fact that I have strategic motives is undoubted, but I also have generous and egalitarian motives, and whenever possible I do things to fulfill them all simultaneously. And when they conflict, many considerations determine which one wins out, but in any given situation my generous or egalitarian motives can in principle win out, as people demonstrate every day as they sacrifice themselves for others.

Some people will also think that our picture of human morality is too rosy because human immorality is on display for all to see all day every day in the world press. All day every day people lie, cheat, and steal to get their own selfish way, and there are at any given moment multiple wars ongoing. But people lying, cheating, and stealing are simply instances when, for whatever reason, the individual's self-interested motives have won out. The liar-cheater-stealer probably felt guilty while she was doing it and attempted to justify it by creative interpretations of the harm (or lack thereof) done. Moreover, she undoubtedly has done many moral things in other situations on other occasions and may even be close to 100 percent moral with family and friends. And as for wars, virtually all large-scale conflicts in the world today are between groups of people who view the situation as "us" versus "them," for example, one country versus another. In addition, there are many other conflicts between different ethnic groups that for various reasons (quite often involving outside influences, e.g., colonialism) have been forced to coexist under the same political umbrella. These are again instances of in-group/out-group conflicts, and again it is almost certain that those involved in them are doing many moral things with their compatriots on a daily basis. And despite all this, it is still the case that warlike conflicts, as well as many other types of violence, are historically on the wane (Pinker, 2011).

A final criticism of too much rosiness is that we have posited a sense of equivalence or equality among persons as foundational to human morality. Those who are used to thinking in terms of recorded human history will point out that it is only with the Enlightenment that social theorists in Western societies began promoting the idea of all individuals as in some sense equal, with equal rights. This is of course true in terms of explicit political thinking about the social contract after the rise of civil societies in the past ten thousand years. But the hunter-gatherer societies that existed for the immediately preceding period—for more than ten times that long—were by all indications highly egalitarian (Boehm, 1999). This does not mean that those individuals had no selfish motives, only that they regularly worked things out with one another in a mutually satisfactory manner based on an equal respect for all members of the cultural group. And recall again that our hypothesis follows the analysis of Nagel (1970) in which the recognition of others as equal beings is not a preference or a motivation (as it is in Enlightenment political treatises) but merely a recognition, perhaps even an unwelcome recognition, that may or may not influence the personal decisions that individuals make or the social norms that cultures create. Indeed, the main way that people justify treating others inhumanely is not motivational but, rather, conceptual: they view them as not really human at all. On the whole, it simply does not seem possible to think of anything resembling human morality without individuals who recognize and interact with others whom they cannot help but perceive as being on a par with themselves.

It is clear that morality is difficult. Human beings have natural inclinations of sympathy and fairness toward others, but still we are sometimes selfish. Others may call us to task for our selfishness, chastise us with the whips of social norms, and gossip behind our backs to ruin our reputation, and still we are sometimes selfish. Violations of our own morality make us feel guilty and chip away at our sense of who we are, and still we are sometimes selfish. Religious principles applied by an omniscient God promise eternal damnation for moral violations, and governmental laws mete out more immediate and concrete forms of damnation in this corporeal world, and still we are sometimes selfish. No, it is a miracle that we are moral, and it did not have to be this way. It just so happens that, on the whole, those of us who made mostly moral decisions most of the time had more babies. And so, again, we should simply marvel and celebrate the fact that, mirabile dictu (and Nietzsche notwithstanding), morality appears to be somehow good for our species, our cultures, and ourselves—at least so far.

Notes

2. Evolution of Cooperation

1. If this sounds familiar, it has indeed been proposed that the evolution of cooperation in social insects is due not to kin selection but to group selection (the homogeneity requirement of group selection means that one is helping individuals genetically very close to oneself)—with bee and ant colonies considered as "superorganisms" (Nowak et al., 2010). Others claim that the two processes are both about positive assortment—cooperators interacting only with cooperators—and so the two approaches are simply mathematical variants of one another.

2. Some theorists argue that the way it works is this: I see you doing something altruistic and infer that you must be someone predisposed to cooperate—and so I interact with you over others. On this basis, you infer the same about me. But from a psychological point of view, this is no kind of reciprocity at all. We are simply choosing one another, not paying one another back.

3. If the opponent plays dove, I can play dove and get half the food or, better, play hawk and get it all. Hawk is better. If the opponent plays hawk, I can play dove and get none of the food or else, better, play hawk and maybe get some or all of it (depending on how the fight goes). Again, hawk is better. So both individuals end up playing hawk, as always in prisoner's dilemma situations.

4. Although there are significant differences in the dominance relations structuring chimpanzee and bonobo societies—chimpanzees being male dominant and bonobos,

by virtue of coalitions, being female dominant—for current purposes these differences are not crucial.

5. This pattern has been well documented not only in great apes but also in savanna baboons, for whom such friendships provide many fitness benefits (Silk et al., 2010; Seyfarth and Cheney, 2012).

6. Horner et al. (2011) used the logic of this design in a different experimental setup and reported some prosocial tendencies in chimpanzees. But the study is uninterpretable because the experimental condition was run before the control condition for all subjects; perhaps subjects were just tired of the prosocial option by the time the control condition came around.

7. The better-known study using this design was run by Brosnan and de Waal (2003) with capuchin monkeys (also without counterbalancing the order of experimental conditions). That study also lacked appropriate control conditions, and failures to replicate using appropriate controls are reported by four other laboratories: Roma et al. (2006), Dubreuil et al. (2006), Fontenot et al. (2007), Sheskin et al. (2013), and McAuliffe et al., 2015.

3. Second-Personal Morality

1. The process would have been significantly different from that practiced by contemporary foragers, as modern foragers have powerful weapons that enable them to hunt individually if they so desire. In addition, contemporary foragers are poor models because they have gone through the second step in our evolutionary story and live in group-minded cultures; thus, for example, even if they forage alone, they often bring back the spoils to share with the group.

2. In general, we cite in this chapter only studies of children three years of age or younger. In this case, however, we have deviated from that rule because no studies of impression management have been conducted with children under five years of age. But even one-year-old infants clearly know when they are being watched and behave differently, most especially by inhibiting their behavior and showing shyness—which other apes do not do (Rochat, 2009).

3. The word *promise* is being avoided here (and this is Scanlon's [1990] actual target), as promises are more public commitments made in a public language—and so we will deal with them in Chapter 4.

4. "Objective" Morality

1. One study suggests that chimpanzees also conform even when they have an already effective method (Whiten et al., 2005), but closer inspection of the data shows that only one individual reliably switched its method of tool use to match that of others.

2. Given a lack of terminological consistency in the literature, we will follow Bicchieri (2006) in using *social norm,* or even *moral norm,* to refer to things that are not

just conventional or traditional ways of doing things (as discussed in the previous section) but, rather, ways of doing things that group members see as necessary for cooperation, with deviations meriting censure.

3. Thus, conventional behaviors are moralized either as individuals who separately have similar evaluative attitudes come to learn that they do (epistemic convergence), or as individuals who do not have this attitude come to follow some others—leaders—who do (attitudinal convergence).

4. Rakoczy and Tomasello (2007) argue that this capacity to collectively create new realities with new deontic powers may be seen already in young children's joint pretense, in which, for example, we agree that this stick is a horse. Wyman et al. (2009) indeed found that when a new agent now treated the stick as merely a stick, preschool children normatively objected by saying things like "no, it's a horse"—"we" have agreed on a new status or identity for the stick.

5. Moral justifications of this type presumably derive from early humans' ways of evaluating others' behavior as cooperators, taking into account their intentions, extenuating circumstances, and the like. That is to say, whereas early humans used these considerations to come to a valid judgment only about a particular act, modern humans began articulating them to influence others' judgments about acts of questionable morality.

6. The contemporary welfare state and its safety net of essentials for life could be seen as subscribing to this same basic philosophy: everyone deserves a minimal amount just for being a member of this one big collaborative activity that is our culture.

References

Alexander, R. D. 1987. *The biology of moral systems.* New York: Aldine De Gruyter.

Alvard, M. 2012. Human social ecology. In J. Mitani, ed., *The evolution of primate societies* (pp. 141–162). Chicago: University of Chicago Press.

Bartal, I., J. Decety, and P. Mason. 2011. Empathy and pro-social behavior in rats. *Science, 334*(6061), 1427–1430.

Batson, C. D. 1991. *The altruism question: Toward a social-psychological answer.* Hillsdale, NJ: Erlbaum.

Baumard, N., J. B. André, and D. Sperber. 2013. A mutualistic approach to morality. *Behavioral and Brain Sciences, 36*(1), 59–122.

Bergson, H. 1935. *Two sources of morality and religion.* New York: Holt.

Bicchieri, C. 2006. *The grammar of society: The nature and dynamics of social norms.* New York: Cambridge University Press.

Bickerton, D., and E. Szathmáry. 2011. Confrontational scavenging as a possible source for language and cooperation. *BMC Evolutionary Biology, 11,* 261.

Blake, P. R., and K. McAuliffe. 2011. "I had so much it didn't seem fair": Eight-year-olds reject two forms of inequity. *Cognition, 120*(2), 215–224.

Blasi, A. 1984. Moral identity: Its role in moral functioning. In W. M. Kurtines and J. J. Gewirtz, eds., *Morality, moral behavior and moral development* (pp. 128–139). New York: Wiley.

Boehm, C. 1999. *Hierarchy in the forest: The evolution of egalitarian behavior.* Cambridge, MA: Harvard University Press.

———. 2012. *Moral origins: The evolution of virtue, altruism, and shame.* New York: Basic Books.

Boesch, C. 1994. Cooperative hunting in wild chimpanzees. *Animal Behaviour, 48*(3), 653–667.

Boesch, C., and H. Boesch. 1989. Hunting behavior of wild chimpanzees in the Taï National Park. *American Journal of Physical Anthropology, 78*(4), 547–573.

Bonnie, K. E., V. Horner, A. Whiten, and F. B. M. de Waal. 2007. Spread of arbitrary conventions among chimpanzees: A controlled experiment. *Proceedings of the Royal Society of London, Series B: Biological Sciences, 274*(1608), 367–372.

Bowles, S., and H. Gintis. 2012. *A cooperative species: Human reciprocity and its evolution.* Princeton, NJ: Princeton University Press.

Boyd, R., and J. Silk. 2009. *How humans evolved.* New York: Norton.

Bratman, M. 1992. Shared co-operative activity. *Philosophical Review, 101*(2), 327–341.

———. 2014. *Shared agency: A planning theory of acting together.* New York: Oxford University Press.

Bräuer, J., J. Call, and M. Tomasello. 2006. Are apes really inequity averse? *Proceedings of the Royal Society of London, Series B: Biological Sciences, 273*(1605), 3123–3128.

———. 2009. Are apes inequity averse? New data on the token-exchange paradigm. *American Journal of Primatology, 71*(2), 175–181.

Brosnan, S. F., and F. B. M. de Waal. 2003. Monkeys reject unequal pay. *Nature, 425,* 297–299.

Brosnan, S. F., T. Flemming, C. F. Talbot, L. Mayo, and T. Stoinski. 2011. Responses to inequity in orangutans. *Folia Primatologica, 82,* 56–70.

Brosnan, S. F., H. C. Schiff, and F. B. M. de Waal. 2005. Tolerance for inequity may increase with social closeness in chimpanzees. *Proceedings of the Royal Society of London, Series B: Biological Sciences,* 272(1560), 253–285.

Brosnan, S. F., C. Talbot, M. Ahlgren, S. P. Lambeth, and S. J. Schapiro. 2010. Mechanisms underlying the response to inequity in chimpanzees, *Pan troglodytes. Animal Behaviour, 79*(6), 1229–1237.

Brown, P., and S. C. Levinson. 1987. *Politeness: Some universals in language usage.* Cambridge: Cambridge University Press.

Bshary, R., and R. Bergmueller. 2008. Distinguishing four fundamental approaches to the evolution of helping. *Journal of Evolutionary Biology, 21*(2), 405–420.

Bullinger, A. F., A. P. Melis, and M. Tomasello. 2011a. Chimpanzees *(Pan troglodytes)* prefer individual over collaborative strategies toward goals. *Animal Behaviour, 82*(5), 1135–1141.

Bullinger, A. F., E. Wyman, A. P. Melis, and M. Tomasello. 2011b. Coordination of chimpanzees *(Pan troglodytes)* in a stag hunt game. *International Journal of Primatology, 32*(6), 1296–1310.

Burkart, J. M., and C. P. van Schaik. 2010. Cognitive consequences of cooperative breeding in primates? *Animal Cognition, 13*(1), 1–19.

Buttelmann, D., J. Call, and M. Tomasello. 2009. Do great apes use emotional expressions to infer desires? *Developmental Science, 12*(5), 688–699.

Buttelmann, D., N. Zmyj, M. M. Daum, and M. Carpenter. 2013. Selective imitation of in-group over out-group members in 14-month-old infants. *Child Development, 84*(2), 422–428.

Call, J., and M. Tomasello. 2007. *The gestural communication of apes and monkeys.* Mahwah, NJ: Erlbaum.

———. 2008. Does the chimpanzee have a theory of mind? 30 years later. *Trends in Cognitive Science, 12*(5), 187–192.

Callaghan, T., H. Moll, H. Rakoczy, F. Warneken, U. Liszkowski, T. Behne, and M. Tomasello. 2011. Early social cognition in three cultural contexts. *Monographs of the Society for Research in Child Development, 76*(2), 1–142.

Carpenter, M. 2006. Instrumental, social, and shared goals and intentions in imitation. In S. J. Rogers and J. Williams, eds., *Imitation and the social mind: Autism and typical development* (pp. 48–70). New York: Guilford Press.

Carpenter, M., M. Tomasello, and T. Striano. 2005. Role reversal imitation in 12 and 18 month olds and children with autism. *Infancy, 8*(3), 253–278.

Carpenter, M., J. Uebel, and M. Tomasello. 2013. Being mimicked increases prosocial behavior in 18-month-old infants. *Child Development, 84*(5), 1511–1518.

Chapais, B. 2008. *Primeval kinship: How pair-bonding gave birth to human society.* Cambridge, MA: Harvard University Press.

Chwe, M. S. Y. 2003. *Rational ritual: Culture, coordination and common knowledge.* Princeton, NJ: Princeton University Press.

Clark, H. 1996. *Using language.* Cambridge: Cambridge University Press.

Clutton-Brock, T. 2002. Breeding together: Kin selection and mutualism in cooperative vertebrates. *Science, 296*(5565), 69–72.

Cosmides, L., and J. Tooby. 2004. Knowing thyself: The evolutionary psychology of moral reasoning and moral sentiments. *Business, Science, and Ethics, 4*, 93–128.

Crockford, C., and C. Boesch. 2003. Context specific calls in wild chimpanzees, *Pan troglodytes verus*: Analysis of barks. *Animal Behaviour, 66*(1), 115–125.

Crockford, C., R. M. Wittig, K. Langergraber, T. E. Ziegler, K. Zuberbühler, and T. Deschner. 2013. Urinary oxytocin and social bonding in related and unrelated wild chimpanzees. *Proceedings of the Royal Society of London, Series B: Biological Sciences, 280*(1755), 2012–2765.

Csibra, G., and G. Gergely. 2009. Natural pedagogy. *Trends in Cognitive Sciences, 13*(4), 148–153.

Darley, J. M., and B. Latane. 1968. Bystander intervention in emergencies: Diffusion of responsibility. *Journal of Personality and Social Psychology, 8*(4), 377–383.

Darwall, S. 1997. Two kinds of respect. *Ethics, 88*, 36–49.

———. 2006. *The second-person standpoint: Respect, morality, and accountability.* Cambridge, MA: Harvard University Press.

———. 2013. *Essays in second-personal ethics,* Vol. 1: *Morality, authority, and law.* Oxford: Oxford University Press.

Darwin, C. 1871. *The descent of man, and selection in relation to sex.* London: John Murray.

Dawkins, R. 1976. *The selfish gene.* New York: Oxford University Press.

de Waal, F. B. M. 1982. *Chimpanzee politics: Power and sex among apes.* London: Cape.

———. 1989a. *Peacemaking among primates.* Cambridge, MA: Harvard University Press.

———. 1989b. Food sharing and reciprocal obligations among chimpanzees. *Journal of Human Evolution, 18*(5), 433–459.

———. 1996. *Good natured: The origins of right and wrong in humans and other animals.* Cambridge, MA: Harvard University Press.

———. 2000. Attitudinal reciprocity in food sharing among brown capuchin monkeys. *Animal Behaviour, 60*(2), 253–261.

———. 2006. *Primates and philosophers: How morality evolved.* Princeton, NJ: Princeton University Press.

de Waal, F. B. M., and L. M. Luttrell. 1988. Mechanisms of social reciprocity in three primate species: Symmetrical relationship characteristics or cognition? *Ethology and Sociobiology, 9*(2–4), 101–118.

Diesendruck, G., N. Carmel, and L. Markson. 2010. Children's sensitivity to the conventionality of sources. *Child Development, 81*(2), 652–668.

Dubreuil, D., M. S. Gentile, and E. Visalberghi. 2006. Are capuchin monkeys *(Cebus apella)* inequality averse? *Proceedings of the Royal Society of London, Series B: Biological Sciences, 273*(1591), 1223–1228.

Duguid, S., E. Wyman, A. Bullinger, and M. Tomasello. 2014. Coordination strategies of chimpanzees and human children in a stag hunt game. *Proceedings of the Royal Society of London, Series B: Biological Sciences, 281,* 20141973.

Dunbar, R. 1998. The social brain hypothesis. *Evolutionary Anthropology, 6*(5), 178–190.

Dunham, Y., A. S. Baron, and M. R. Banaji. 2008. The development of implicit intergroup cognition. *Trends in Cognitive Sciences, 12*(7), 248–253.

Durkheim, E. 1893/1984. *The division of labor in society.* New York: Free Press.

———. 1912/2001. *The elementary forms of religious life.* Oxford: Oxford University Press.

———. 1974. *Sociology and philosophy.* New York: Free Press.

Engelmann, J. M., E. Herrmann, and M. Tomasello. 2012. Five-year olds, but not chimpanzees, attempt to manage their reputations. *PLoS ONE, 7*(10), e48433.

Engelmann, J., E. Herrmann, and M. Tomasello. In press. Chimpanzees trust conspecifics to engage in low-cost reciprocity. *Proceedings of the Royal Society B.*

———. Submitted. Young children overcome peer pressure to do the right thing.

Engelmann, J. M., H. Over, E. Herrmann, and M. Tomasello. 2013. Young children care more about their reputation with ingroup members and potential reciprocators. *Developmental Science, 16*(6), 952–958.

Fehr, E., H. Bernhard, and B. Rockenbach. 2008. Egalitarianism in young children. *Nature, 454*, 1079–1083.

Fiske, S. T. 2010. *Social beings: Core motives in social psychology.* 2nd ed. Hoboken, NJ: Wiley.

Fletcher, G., F. Warneken, and M. Tomasello. 2012. Differences in cognitive processes underlying the collaborative activities of children and chimpanzees. *Cognitive Development, 27*(2), 136–153.

Foley, R. A., and C. Gamble. 2009. The ecology of social transitions in human evolution. *Philosophical Transactions of the Royal Society of London, Series B: Biological Sciences, 364*(1442), 3267–3279.

Fontenot, M. B., S. L. Watson, K. A. Roberts, and R. W. Miller. 2007. Effects of food preferences on token exchange and behavioural responses to inequality in tufted capuchin monkeys, *Cebus apella. Animal Behaviour, 74*(3), 487–496.

Friedrich, D., and N. Southwood. 2011. Promises and trust. In H. Sheinman, ed., *Promises and agreement: Philosophical essays* (pp. 275–292). New York: Oxford University Press.

Gibbard, A. 1990. *Wise choices, apt feelings: A theory of normative judgment.* Cambridge, MA: Harvard University Press.

Gilbert, M. 1990. Walking together: A paradigmatic social phenomenon. *Midwest Studies in Philosophy, 15*(1), 1–14.

———. 2003. The structure of the social atom: Joint commitment as the foundation of human social behavior. In F. Schmitt, ed., *Socializing metaphysics* (pp. 39–64). Lanham, MD: Rowman and Littlefield.

———. 2006. *A theory of political obligation: Membership, commitment, and the bonds of society.* Oxford: Oxford University Press.

———. 2011. Three dogmas about promising. In H. Sheinman, ed., *Promises and agreements* (pp. 80–109). New York: Oxford University Press.

———. 2014. *Joint commitment: How we make the social world.* New York: Oxford University Press.

Gilby, I. C. 2006. Meat sharing among the Gombe chimpanzees: Harassment and reciprocal exchange. *Animal Behaviour, 71*(4), 953–963.

Göckeritz, S., M. F. H. Schmidt, and M. Tomasello. 2014. Young children's creation and transmission of social norms. *Cognitive Development, 30*(April–June), 81–95.

———. Submitted. Young children understand norms as socially constructed—if they have done the constructing.

Goffman, E. 1959. *The presentation of self in everyday life.* New York: Anchor.

Gomes, C., C. Boesch, and R. Mundry. 2009. Long-term reciprocation of grooming in wild West African chimpanzees. *Proceedings of the Royal Society of London, Series B: Biological Sciences, 276,* 699–706.

Goodall, J. 1986. *The chimpanzees of Gombe: Patterns of behavior.* Cambridge, MA: Belknap Press.

Gräfenhain, M., T. Behne, M. Carpenter, and M. Tomasello. 2009. Young children's understanding of joint commitments. *Developmental Psychology, 45*(5), 1430–1443.

Gräfenhain, M., M. Carpenter, and M. Tomasello. 2013. Three-year-olds' understanding of the consequences of joint commitments. *PLoS ONE, 8*(9), e73039.

Greenberg, J. R., K. Hamann, F. Warneken, and M. Tomasello. 2010. Chimpanzee helping in collaborative and non-collaborative contexts. *Animal Behaviour, 80*(5), 873–880.

Greene, J. 2013. *Moral tribes: Emotion, reason, and the gap between us and them.* New York: Penguin Press.

Greene, J. D., R. B. Sommerville, L. E. Nystrom, J. M. Darley, and J. D. Cohen. 2001. An fMRI investigation of emotional engagement in moral judgment. *Science, 293*(5537), 2105–2108.

Grocke, P., F. Rossano, and M. Tomasello. In press. Preschoolers accept unequal resource distributions if the procedure provides equal opportunities. *Journal of Experimental Child Psychology.*

Grüneisen, S., E. Wyman, and M. Tomasello. 2015. Conforming to coordinate: Children use majority information for peer coordination. *British Journal of Developmental Psychology, 33*(1), 136–147.

Guererk, O., B. Irlenbusch, and B. Rockenbach. 2006. The competitive advantage of sanctioning institutions. *Science, 312*(5770), 108–111.

Gurven, M. 2004. To give or not to give: An evolutionary ecology of human food transfers. *Behavioral and Brain Sciences, 27*(4), 543–583.

Haidt, J. 2012. *The righteous mind: Why good people are divided by politics and religion.* New York: Pantheon.

Haley, K. J., and D. M. T. Fessler. 2005. Nobody's watching? Subtle cues affect generosity in an anonymous economic game. *Evolution and Human Behavior, 26*(3), 245–256.

Hamann, K., J. Bender, and M. Tomasello. 2014. Meritocratic sharing is based on collaboration in 3-year-olds. *Developmental Psychology, 50*(1), 121–128.

Hamann, K., F. Warneken, J. Greenberg, and M. Tomasello. 2011. Collaboration encourages equal sharing in children but not chimpanzees. *Nature, 476,* 328–331.

Hamann, K., F. Warneken, and M. Tomasello. 2012. Children's developing commitments to joint goals. *Child Development, 83*(1), 137–145.

Hamlin, J. K., K. Wynn, and P. Bloom. 2007. Social evaluation by preverbal infants. *Nature, 450,* 557–559.

Harcourt, A. H., and F. B. M. de Waal, eds. 1992. *Coalitions and alliances in humans and other animals*. Oxford: Oxford University Press.

Hardy, S. A., and G. Carlo. 2005. Identity as a source of moral motivation. *Human Development, 48*(4), 232–256.

Hare, B. 2001. Can competitive paradigms increase the validity of social cognitive experiments in primates? *Animal Cognition, 4*(3–4), 269–280.

Hare, B., J. Call, B. Agnetta, and M. Tomasello. 2000. Chimpanzees know what conspecifics do and do not see. *Animal Behaviour, 59*(4), 771–785.

Hare, B., J. Call, and M. Tomasello. 2001. Do chimpanzees know what conspecifics know and do not know? *Animal Behaviour, 61*(1), 139–151.

Hare, B., and M. Tomasello. 2004. Chimpanzees are more skillful in competitive than in cooperative cognitive tasks. *Animal Behaviour, 68*(3), 571–581.

Hare, B., T. Wobber, and R. Wrangham. 2012. The self-domestication hypothesis: Bonobo psychology evolved due to selection against male aggression. *Animal Behavior, 83*, 573–585.

Haun, D., and H. Over. 2014. Like me: A homophily-based account of human culture. In P. J. Richerson and M. Christiansen, eds., *Cultural evolution* (pp. 75–85). Cambridge, MA: MIT Press.

Haun, D. B. M., and M. Tomasello. 2011. Conformity to peer pressure in preschool children. *Child Development, 82*(6), 1759–1767.

———. 2014. Great apes stick with what they know; children conform to others. *Psychological Science, 25*(12), 2160–2167.

Hegel, G. W. F. 1807/1967. *The phenomenology of mind*. New York: Harper and Row.

Henrich, J., R. Boyd, S. Bowles, C. Camerer, H. Gintis, R. McElreath, and E. Fehr. 2001. In search of *Homo economicus:* Experiments in 15 small-scale societies. *American Economic Review, 91*(2), 73–79.

Hepach, R., A. Vaish, and M. Tomasello. 2012. Young children are intrinsically motivated to see others helped. *Psychological Science, 23*(9), 967–972.

———. 2013. Young children sympathize less in response to unjustified emotional distress. *Developmental Psychology, 49*(6), 1132–1138.

Herrmann, E., S. Keupp, B. Hare, A. Vaish, and M. Tomasello. 2013. Direct and indirect reputation formation in non-human great apes and human children. *Journal of Comparative Psychology, 127*(1), 63–75.

Herrmann , E., A. Misch, and M. Tomasello. In press. Uniquely human self-control begins at school age. *Developmental Science*, doi: 10.1111/desc.12272.

Hill, K. 2002. Altruistic cooperation during foraging by the Ache, and the evolved human predisposition to cooperate. *Human Nature, 13*(1), 105–128.

———. 2009. The emergence of human uniqueness: Characteristics underlying behavioural modernity. *Evolutionary Anthropology, 18*, 187–200.

Hill, K., M. Barton, and A. M. Hurtado. 2009. The emergence of human uniqueness: Characters underlying behavioral modernity. *Evolutionary Anthropology, 18*(5), 187–200.

Hoffman, M. L. 2000. *Empathy and moral development: Implications for caring and justice.* Cambridge: Cambridge University Press.

Honneth, A. 1995. *The struggle for recognition: The moral grammar of social conflicts.* Cambridge: Polity Press.

Hopper, L. M., S. P. Lambeth, S. J. Schapiro, and S. F. Brosnan. 2013. When given the opportunity, chimpanzees maximize personal gain rather than "level the playing field." *PeerJ, 1,* e165.

Horner, V., J. D. Carter, M. Suchak, and F. B. M. de Waal. 2011. Spontaneous prosocial choice by chimpanzees. *Proceedings of the National Academy of Sciences of the United States of America, 108*(33), 13847–13851.

Horner, V., and A. K. Whiten. 2005. Causal knowledge and imitation/emulation switching in chimpanzees *(Pan troglodytes)* and children. *Animal Cognition, 8*(3), 164–181.

House, B. R., J. B. Silk, J. Henrich, H. C. Barrett, B. A. Scelza, A. H. Boyette, B. S. Hewlett, R. McElreath, and S. Laurence. 2013. Ontogeny of prosocial behavior across diverse societies. *Proceedings of the National Academy of Sciences of the United States of America, 110*(36), 14586–14591.

Hrdy, S. 2009. *Mothers and others: The evolutionary origins of mutual understanding.* Cambridge, MA: Belknap Press.

Hruschka, D. J. 2010. *Friendship: Development, ecology and evolution of a social relationship.* Berkeley, CA: University of California Press.

Hume, D. 1751/1957. *An enquiry concerning the principles of morals.* New York: Bobbs-Merrill.

Jensen, K., J. Call, and M. Tomasello. 2007. Chimpanzees are rational maximizers in an ultimatum game. *Science, 318*(5847), 107–109.

Jensen, K., B. Hare, J. Call, and M. Tomasello. 2006. What's in it for me? Self-regard precludes altruism and spite in chimpanzees. *Proceedings of the Royal Society of London, Series B: Biological Sciences, 273*(1589), 1013–1021.

Jensen, K., and J. B. Silk. 2014. Searching for the evolutionary roots of human morality. In M. Killen and J. G. Smetana, eds., *Handbook of moral development* (2nd ed., pp. 475–494). New York: Psychology Press.

Joyce, R. 2006. *The evolution of morality.* Cambridge, MA: MIT Press.

Kaiser, I., K. Jensen, J. Call, and M. Tomasello. 2012. Theft in an ultimatum game: Chimpanzees and bonobos are insensitive to unfairness. *Biology Letters, 8*(6), 942–945.

Kanngiesser, P., and F. Warneken. 2012. Young children take merit into account when sharing rewards. *PLoS ONE, 7*(8), e43979.

Kant, I. 1785/1988. *Fundamental principles of the metaphysics of morals.* Buffalo, NY: Prometheus.

Killen, M., K. L. Mulvey, and A. Hitti. 2013. Social exclusion in childhood: A developmental intergroup perspective. *Child Development, 84*(3), 772–790.

Kinzler, K. D., K. H. Corriveau, and P. L Harris. 2011. Children's selective trust in native-accented speakers. *Developmental Science, 14*(1), 106–111.

Kinzler, K. D., K. Shutts, J. DeJesus, and E. S. Spelke. 2009. Accent trumps race in guiding children's social preferences. *Social Cognition, 27*(4), 623–634.

Kirschner, S., and M. Tomasello. 2010. Joint music making promotes prosocial behavior in 4-year-old children. *Evolution and Human Behavior, 31*(5), 354–364.

Kitcher, P. 2011. *The ethical project.* Cambridge, MA: Harvard University Press.

Klein, R. 2009. *The human career: Human biological and cultural origins.* 3rd ed. Chicago: University of Chicago Press.

Knight, J. 1992. *Institutions and social conflict.* Cambridge, MA: Cambridge University Press.

Kohlberg, L. 1981. *Essays on moral development,* Vol. 1: *The philosophy of moral development.* San Francisco, CA: Harper and Row.

Kojève, A. 1982/2000. *Outline of a phenomenology of right.* New York: Roman and Littlefield.

Korsgaard, C. 1996a. *The sources of normativity.* Cambridge: Cambridge University Press.

———. 1996b. *Creating the kingdom of ends.* Cambridge: Cambridge University Press.

Koski, S. E., H. de Vries, S. W. van den Tweel, and E. H. M. Sterck. 2007. What to do after a fight? The determinants and inter-dependency of post-conflict interactions in chimpanzees. *Behaviour, 144*(5), 529–555.

Köymen, B., E. Lieven, D. A. Engemann, H. Rakoczy, F. Warneken, and M. Tomasello. 2014. Children's norm enforcement in their interactions with peers. *Child Development, 85*(3), 1108–1122.

Köymen, B., M. F. H. Schmidt, and M. Tomasello. In press. Teaching versus enforcing norms in preschoolers' peer interactions. *Journal of Experimental Child Psychology.*

Kropotkin, P. A. 1902. *Mutual aid: A factor of evolution.* New York: McClure, Philips.

Kruger, A., and M. Tomasello. 1996. Cultural learning and learning culture. In D. Olson, ed., *Handbook of education and human development: New models of learning, teaching, and schooling* (pp. 369–387). Cambridge, MA: Blackwell.

Kuhlmeier, V., K. Wynn, and P. Bloom. 2003. Attribution of dispositional states by 12-month-olds. *Psychological Science, 14*(5), 402–408.

Kummer, H. (1979). On the value of social relationships to nonhuman primates: A heuristic scheme. In M. von Cranach, K. Foppa, W. Lepenies, and D. Ploog, eds., *Human ethology: Claims and limits of a new discipline* (pp. 381–395). Cambridge: Cambridge University Press.

Lakatos, I., and A. Musgrave, eds. 1970. *Criticism and the growth of knowledge.* Cambridge: Cambridge University Press.

Langergraber, K. E., J. C. Mitani, and L. Vigilant. 2007. The limited impact of kinship on cooperation in wild chimpanzees. *Proceedings of the National Academy of Sciences of the United States of America, 104*(19), 7786–7790.

Leach, H. M. 2003. Human domestication reconsidered. *Current Anthropology, 44*(3), 349–368.

Levinson, S. 2006. On the human interactional engine. In N. Enfield and S. Levinson, eds., *Roots of human sociality* (pp. 39–69). New York: Berg.

Lewis, D. 1969. *Convention: A philosophical study.* Cambridge, MA: Harvard University Press.

Lickel, B., T. Schmader, and M. Spanovic. 2007. Group-conscious emotions: The implications of others' wrongdoings for identity and relationships. In J. L. Tracy, R. W. Robins, and J. P. Tangney, eds., *The self-conscious emotions: Theory and research* (pp. 351–369). New York: Guilford Press.

Liebal, K., M. Carpenter, and M. Tomasello. 2013. Young children's understanding of cultural common ground. *British Journal of Developmental Psychology, 31*(1), 88–96.

Liebal, K., A. Vaish, D. Haun, and M. Tomasello. 2014. Does sympathy motivate prosocial behavior in great apes? *PLoS ONE, 9*(1), e84299.

List, C., and P. Pettit. 2011. *Group agency.* Oxford University Press.

Mameli, M. 2013. Meat made us moral: A hypothesis on the nature and evolution of moral judgment. *Biological Philosophy, 28,* 903–931.

Marlowe, F. W., and J. C. Berbesque. 2008. More "altruistic" punishment in larger societies. *Proceedings of the Royal Society of London, Series B: Biological Sciences, 275*(1634), 587–590.

Martin, A., and K. R. Olson. 2013. When kids know better: Paternalistic helping in 3-year-old children. *Developmental Psychology, 49*(11), 2071–2081.

Marx, K. 1867/1977. *Capital: A critique of political economy.* Vol. 1. New York: Vintage Books.

Maynard Smith, J. 1982. *Evolution and the theory of games.* Cambridge: Cambridge University Press.

McAuliffe, K., J. Jordan, and F. Warneken. 2015. Costly third-party punishment in young children. *Cognition, 134,* 1–10.

Mead, G. H. 1934. *Mind, self, and society. From the standpoint of a social behaviorist.* Chicago: University of Chicago Press.

Melis, A. P., K. Altricher, and M. Tomasello. 2013. Allocation of resources to collaborators and free-riders by 3-year-olds. *Journal of Experimental Child Psychology, 114*(2), 364–370.

Melis, A. P., B. Hare, and M. Tomasello. 2006a. Chimpanzees recruit the best collaborators. *Science, 311*(5765), 1297–1300.

———. 2006b. Engineering cooperation in chimpanzees: Tolerance constraints on cooperation. *Animal Behaviour, 72*(2), 275–286.

———. 2008. Do chimpanzees reciprocate received favors? *Animal Behaviour, 76*(3), 951–962.

———. 2009. Chimpanzees coordinate in a negotiation game. *Evolution and Human Behavior, 30*(6), 381–392.

Melis, A. P., A.-C. Schneider, and M. Tomasello. 2011a. Chimpanzees share food in the same way after collaborative and individual food acquisition. *Animal Behaviour, 82*(3), 485–493.

Melis, A. P., and M. Tomasello. 2013. Chimpanzees' strategic helping in a collaborative task. *Biology Letters, 9,* 20130009.

Melis, A. P., F. Warneken, K. Jensen, A.-C. Schneider, J. Call, and M. Tomasello. 2011b. Chimpanzees help conspecifics to obtain food and non-food items. *Proceedings of the Royal Society of London, Series B: Biological Sciences, 278*(1710), 1405–1413.

Mikhail, J. 2007. Universal moral grammar: Theory, evidence and the future. *Trends in Cognitive Sciences, 11*(4), 143–152.

Milinski, M., D. Semmann, and H.-J. Krambeck. 2002. Reputation helps solve the "tragedy of the commons." *Nature, 415,* 424–426.

Miller, J. G. 1994. Cultural diversity in the morality of caring: Individually oriented versus duty-based interpersonal moral codes. *Cross Cultural Research, 28*(1), 3–39.

Millikan, R. G. 2005. *Language: A biological model.* Oxford: Oxford University Press.

Misch, A., H. Over, and M. Carpenter. 2014. Stick with your group: Young children's attitudes about group loyalty. *Journal of Experimental Child Psychology, 126,* 19–36.

Mitani, J. C., and D. Watts. 2001. Why do chimpanzees hunt and share meat? *Animal Behaviour, 61*(5), 915–924.

Moll, H., N. Richter, M. Carpenter, and M. Tomasello. 2008. Fourteen-month-olds know what "we" have shared in a special way. *Infancy, 13,* 90–101.

Moll, H., and M. Tomasello. 2007a. Co-operation and human cognition: The Vygotskian intelligence hypothesis. *Philosophical Transactions of the Royal Society of London, Series B: Biological Sciences, 362*(1480), 639–648.

———. 2007b. How 14- and 18-month-olds know what others have experienced. *Developmental Psychology, 43*(2), 309–317.

Moreland, R. L. 2010. Are dyads really groups? *Small Group Research, 41*(2), 251–267.

Mueller, M. N., and J. C. Mitani. 2005. Conflict and cooperation in wild chimpanzees. *Advances in the Study of Behavior, 35,* 275–331.

Mussweiler, T. 2003. Comparison processes in social judgment: Mechanisms and consequences. *Psychological Review, 110*(3), 472–489.

Nagel, T. 1970. *The possibility of altruism.* Princeton, NJ: Princeton University Press.

———. 1986. *The view from nowhere.* New York: Oxford University Press.

———. 1991. *Equality and partiality.* New York: Oxford University Press.

Nichols, S. 2004. *Sentimental rules: On the natural foundations of moral judgment.* Oxford: Oxford University Press.

Nichols, S., M. Svetlova, and C. Brownell. 2009. The role of social understanding and empathic disposition in young children's responsiveness to distress in parents and peers. *Cognition, Brain, Behavior, 4*, 449–478.

Nietzsche, F. 1887/2003. *The genealogy of morals.* Mineola, NY: Dover.

Noe, R., and Hammerstein, P. 1994. Biological markets: Supply and demand determine the effect of partner choice in cooperation, mutualism and mating. *Behavioral Ecology and Sociobiology, 35*(1), 1–11.

Norenzayan, A. 2013. *Big gods: How religion transformed cooperation and conflict.* Princeton, NJ: Princeton University Press.

Nowak, M., and R. Highfield. 2011. *Supercooperators: The mathematics of evolution, altruism and human behaviour.* Edinburgh: Canongate.

Nowak, M., C. Tarnita, and E. Wilson. 2010. The evolution of eusociality. *Nature, 466*, 1057–1062.

Olson, K. R., and E. S. Spelke. 2008. Foundations of cooperation in young children. *Cognition, 108*(1), 222–231.

Olson, M. 1965. *The logic of collective action.* Cambridge, MA: Harvard University Press.

Over, H., and M. Carpenter. 2013. The social side of imitation. *Child Development Perspectives, 7*(1), 6–11.

Over, H., M. Carpenter, R. Spears, and M. Gattis. 2013. Children selectively trust individuals who have imitated them. *Social Development, 22*(2), 215–425.

Over, H., A. Vaish, and M. Tomasello. Submitted. Young children accept responsibility for the negative actions of in-group members.

Piaget, J. 1932/1997. *The moral judgment of the child.* New York: Free Press.

Pinker, S. 2011. *The better angels of our nature: Why violence has declined.* New York: Viking.

Prinz, J. 2007. *The emotional construction of morals.* Oxford: Oxford University Press.

———. 2012. *Beyond human nature.* London: Allen Lane.

Proctor, D., R. A. Williamson, F. B. M. de Waal, and S. F. Brosnan. 2013. Chimpanzees play the ultimatum game. *Proceedings of the National Academy of Sciences of the United States of America, 110*(6), 2070–2075.

Rakoczy, H., K. Hamann, F. Warneken, and M. Tomasello. 2010. Bigger knows better? Young children selectively learn rule games from adults rather than from peers. *British Journal of Developmental Psychology, 28*(4), 785–798.

Rakoczy, H., and M. Tomasello. 2007. The ontogeny of social ontology: Steps to shared intentionality and status functions. In S. Tsohatzidis, ed., *Intentional acts and institutional facts: Essays on John Searle's social ontology* (pp. 113–137). Dordrecht: Springer.

Rakoczy, H., F. Warneken, and M. Tomasello. 2008. The sources of normativity: Young children's awareness of the normative structure of games. *Developmental Psychology, 44*(3), 875–881.

Rand, D., J. Greene, and M. A. Nowak. 2012. Spontaneous giving and calculated greed. *Nature, 489*, 427–430.

Rawls, J. 1971. *A theory of justice.* Cambridge, MA: Harvard University Press.

Rekers, Y., D. B. M. Haun, and M. Tomasello. 2011. Children, but not chimpanzees, prefer to collaborate. *Current Biology, 21*(20), 1756–1758.

Resnick, P., R. Zeckhauser, J. Swanson, and K. Lockwood. 2006. The value of reputation on eBay: A controlled experiment. *Experimental Economy, 9*(2), 79–101.

Rhodes, M., and L. Chalik. 2013. Social categories as markers of intrinsic interpersonal obligations. *Psychological Science, 24*(6), 999–1006.

Richerson, P., and R. Boyd. 2005. *Not by genes alone: How culture transformed human evolution.* Chicago: University of Chicago Press.

Riedl, K., K. Jensen, J. Call, and M. Tomasello. 2012. No third-party punishment in chimpanzees. *Proceedings of the National Academy of Sciences of the United States of America, 109*(37), 14824–14829.

———. 2015. Restorative justice in children. *Current Biology, 25*, 1–5.

Roberts, G. 2005. Cooperation through interdependence. *Animal Behaviour, 70*(4), 901–908.

Rochat, P. 2009. *Others in mind: Social origins of self-consciousness.* Cambridge: Cambridge University Press.

Rochat, P., M. D. G. Dias, L. Guo, T. Broesch, C. Passos-Ferreira, A. Winning, and B. Berg. 2009. Fairness in distributive justice by 3- and 5-year-olds across 7 cultures. *Journal of Cross-Cultural Psychology, 40*(3), 416–442.

Rockenbach, B., and M. Milinski. 2006. The efficient interaction of indirect reciprocity and costly punishment. *Nature, 444*, 718–723.

Roma, P. G., A. Silberberg, A. M. Ruggiero, and S. J. Suomi. 2006. Capuchin monkeys, inequity aversion, and the frustration effect. *Journal of Comparative Psychology, 120*(1), 67–73.

Rose, L. M., S. Perry, M. Panger, K. Jack, J. Manson, J. Gros-Luis, K. C. Mackinnon, and E. Vogel. 2003. Interspecific interactions between *Cebus capuchinus* and other species in Costa Rican sites. *International Journal of Primatology, 24*(4), 759–796.

Rossano, F., H. Rakoczy, and M. Tomasello. 2011. Young children's understanding of violations of property rights. *Cognition, 123*(2), 219–227.

Roughley, N. 2015. Resentment, empathy and moral normativity. In N. Roughley and T. Schramme, eds., *Forms of fellow feeling: Sympathy, empathy, concern and moral agency* (pp. 225–247). Cambridge: Cambridge University Press.

Rousseau, J. J. 1762/1968. *Of the social contract, or, Principles of political right.* New York: Penguin.

Scanlon, T. M. 1990. Promises and practices. *Philosophy and Public Affairs, 19*(3), 199–226.

———. 1998. *What we owe to each other.* Cambridge, MA: Belknap Press.

Schäfer, M., D. Haun, and M. Tomasello. In press. Fair is not fair everywhere. *Psychological Science.*

Schino, G., and F. Aureli. 2009. Reciprocal altruism in primates: Partner choice, cognition and emotions. *Advances in the Study of Behavior, 39,* 45–69.

Schmidt, M. F. H., H. Rakoczy, and M. Tomasello. 2012. Young children enforce social norms selectively depending on the violator's group affiliation. *Cognition, 124*(3), 325–333.

———. 2013. Young children understand and defend the entitlements of others. *Journal of Experimental Child Psychology, 116*(4), 930–944.

Schmidt, M. F. H., and M. Tomasello. 2012. Young children enforce social norms. *Current Directions in Psychological Science, 21*(4), 232–236.

Searle, J. 1995. *The construction of social reality.* New York: Free Press.

———. 2010. *Making the social world: The structure of human civilization.* New York: Oxford University Press.

Sellars, W. 1963. *Science, perception and reality.* New York: Humanities Press.

Seyfarth, R. M., and D. L. Cheney. 2012. The evolutionary origins of friendship. *Annual Review of Psychology, 63,* 153–177.

Shapiro, S. 2011. *Legality.* Cambridge, MA: Harvard University Press.

Shaw, A., and K. R. Olson. 2012. Children discard a resource to avoid inequality. *Journal of Experimental Psychology: General, 141*(2), 382–395.

———. 2014. Fairness as partiality aversion: The development of procedural justice. *Journal of Experimental Child Psychology, 119,* 40–53.

Sheskin, M., K. Ashayeri, A. Skerry, and L. R. Santos. 2013. Capuchin monkeys *(Cebus apella)* fail to show inequality aversion in a no-cost situation. *Evolution and Human Behavior, 35*(2), 80–88.

Shweder, R. A., M. Mahapatra, and J. G. Miller. 1987. Culture and moral development. In J. Kagan and S. Lamb, eds., *The emergence of morality in young children* (pp. 1–83). Chicago: University of Chicago Press.

Silk, J. B. 2009. Nepotistic cooperation in nonhuman primate groups. *Philosophical Transactions of the Royal Society of London, Series B: Biological Sciences, 364*(1533), 3243–3254.

Silk, J. B., J. C. Beehner, T. J. Berman, C. Crockford, A. L. Engh, L. R. Moscovice, R. M. Wittig, R. M. Seyfarth, and D. L. Cheney. 2010. Strong and consistent social bonds enhance the longevity of female baboons. *Current Biology, 20*(15), 1359–1361.

Silk, J. B., S. F. Brosnan, J. Vonk, J. Henrich, D. J. Povinelli, A. F. Richardson, S. P. Lambeth, J. Mascaro, and S. J. Schapiro. 2005. Chimpanzees are indifferent to the welfare of other group members. *Nature, 435,* 1357–1359.

Simmel, G. 1908. *Sociology: Investigations on the forms of sociation.* Leipzig: Duncker and Humblot.

Sinnott-Armstrong, W., and T. Wheatley. 2012. The disunity of morality and why it matters to philosophy. *Monist, 95,* 355–377.

Skyrms, B. 2004. *The stag hunt and the evolution of sociality.* Cambridge: Cambridge University Press.

Smith, A. 1759/1982. *The theory of moral sentiments.* Indianapolis, IN: Liberty Classics.

Smith, A. M. 2013. Moral blame and moral protest. In D. J. Coates and N. A. Tognazzini, eds., *Blame: Its nature and norms* (pp. 27–48). New York: Oxford University Press.

Smith, C., P. R. Blake, and P. L. Harris. 2013. I should but I won't: Why young children endorse norms of fair sharing but do not follow them. *PloS ONE, 8*(8), e59510.

Sober, E., and D. S. Wilson. 1998. *Unto others: The evolution and psychology of unselfish behavior.* Cambridge, MA: Harvard University Press.

Steadman, L. B., C. T. Palmer, and C. Tilley. 1996. The universality of ancestor worship. *Ethnology, 35*(1), 63–76.

Sterelny, K. 2012. *The evolved apprentice.* Cambridge, MA: MIT Press.

Stiner, M. 2013. An unshakable Middle Paleolithic? Trends versus conservatism in the predatory niche and their social ramifications. *Current Anthropology, 54*(Suppl. 8), 288–304.

Strawson, P. F. 1962. Freedom and resentment. *Proceedings of the British Academy, 48,* 1–25.

Surbeck, M., and G. Hohmann. 2008. Primate hunting by bonobo at LuiKotale, Salonga National Park. *Current Biology, 18*(19), R906–R907.

Svetlova, M., S. Nichols, and C. Brownell. 2010. Toddlers' prosocial behavior: From instrumental to empathic to altruistic helping. *Child Development, 81*(6), 1814–1827.

Sylwester, K., and G. Roberts. 2010. Cooperators benefit through reputation-based partner choice in economic games. *Biological Letters, 6,* 659–662.

Tangney, J. P., and R. L. Dearing. 2004. *Shame and guilt.* New York: Guilford Press.

Tennie, C., J. Call, and M. Tomasello. 2009. Ratcheting up the ratchet: On the evolution of cumulative culture. *Philosophical Transactions of the Royal Society of London, Series B: Biological Sciences, 364*(1528), 2405–2415.

Thompson, M. 2008. *Life and action: Elementary structures of practice and practical thought.* Cambridge, MA: Harvard University Press.

Thornton, A., and N. J. Raihani. 2008. The evolution of teaching. *Animal Behaviour, 75*(6), 1823–1836.

Tomasello, M. 1995. Joint attention as social cognition. In C. Moore and P. Dunham, eds., *Joint attention: Its origins and role in development* (pp. 103–130). Hillsdale, NJ: Erlbaum.

———. 2006. Conventions are shared (commentary on Millikan, *Language: A biological model*). *Philosophy of Mind Review, 5*, 29–36.

———. 2008. *Origins of human communication.* Cambridge, MA: MIT Press.

———. 2009. *Why we co-operate.* Cambridge, MA: MIT Press.

———. 2011. Human culture in evolutionary perspective. In M. Gelfand, ed., *Advances in culture and psychology* (pp. 3–31). Oxford: Oxford University Press.

———. 2014. *A natural history of human thinking.* Cambridge, MA: Harvard University Press.

Tomasello, M., and M. Carpenter. 2005. The emergence of social cognition in three young chimpanzees. *Monographs of the Society for Research in Child Development, 70*(1), 1–152.

Tomasello, M., M. Carpenter, J. Call, T. Behne, and H. Moll. 2005. Understanding and sharing intentions: The origins of cultural cognition. *Behavioral and Brain Sciences, 28*(5), 675–691.

Tomasello, M., A. P. Melis, C. Tennie, E. Wyman, and E. Herrmann. 2012. Two key steps in the evolution of cooperation: The interdependence hypothesis. *Current Anthropology, 53*(6), 673–692.

Tomasello, M., and A. Vaish. 2013. Origins of human cooperation and morality. *Annual Review of Psychology, 64*, 231–255.

Trivers, R. 1971. The evolution of reciprocal altruism. *Quarterly Review of Biology, 46*(1), 35–57.

Tuomela, R. 2007. *The philosophy of sociality: The shared point of view.* Oxford: Oxford University Press.

Turiel, E. 1983. *The development of social knowledge: Morality and convention.* Cambridge, MA: Cambridge University Press.

———. 2006. The development of morality. In W. Damon and R. M. Lerner, eds., *Handbook of child psychology*, Vol. 3: *Social, emotional, and personality development* (pp. 253–300). New York: Wiley.

Ulber, J., K. Hamann, and M. Tomasello. Submitted. Division behavior in 18- and 24-month-old peers.

Vaish, A., M. Carpenter, and M. Tomasello. 2009. Sympathy through affective perspective-taking and its relation to prosocial behavior in toddlers. *Developmental Psychology, 45*(2), 534–543.

———. 2010. Young children selectively avoid helping people with harmful intentions. *Child Development, 81*(6), 1661–1669.

———. 2011a. Young children's responses to guilt displays. *Developmental Psychology, 47*(5), 1248–1262.

———. In press. Three-year-olds feel guilt only when appropriate. *Child Development.*

Vaish, A., E. Herrmann, C. Markmann, and M. Tomasello. Submitted. Three-year-olds prefer norm enforcers.

Vaish, A., M. Missana, and M. Tomasello. 2011b. Three-year-old children intervene in third-party moral transgressions. *British Journal of Developmental Psychology, 29*(1), 124–130.

von Rohr, C., J. Burkart, and C. van Schaik. 2011. Evolutionary precursors of social norms in chimpanzees: A new approach. *Biology and Philosophy, 26,* 1–30.

Vygotsky, L. 1978. *Mind in society: The development of higher psychological processes.* Edited by M. Cole. Cambridge, MA: Harvard University Press.

Warneken, F. 2013. Young children proactively remedy unnoticed accidents. *Cognition, 126*(1), 101–108.

Warneken, F., F. Chen, and M. Tomasello. 2006. Cooperative activities in young children and chimpanzees. *Child Development, 77*(3), 640–663.

Warneken, F., B. Hare, A. Melis, D. Hanus, and M. Tomasello. 2007. Spontaneous altruism by chimpanzees and young children. *PLoS Biology, 5*(7), e184.

Warneken, F., K. Lohse, A. P. Melis, and M. Tomasello. 2011. Young children share the spoils after collaboration. *Psychological Science, 22*(2), 267–273.

Warneken, F., and M. Tomasello. 2006. Altruistic helping in human infants and young chimpanzees. *Science, 311*(5765), 1301–1303.

———. 2007. Helping and cooperation at 14 months of age. *Infancy, 11*(3), 271–294.

———. 2008. Extrinsic rewards undermine altruistic tendencies in 20-month-olds. *Developmental Psychology, 44*(6), 1785–1788.

———. 2009. Varieties of altruism in children and chimpanzees. *Trends in Cognitive Science, 13*(9), 397–402.

———. 2013. The emergence of contingent reciprocity in young children. *Journal of Experimental Child Psychology, 116*(2), 338–350.

Watts, D., and J. C. Mitani. 2002. Hunting behavior of chimpanzees at Ngogo, Kibale National Park, Uganda. *International Journal of Primatology, 23*(1), 1–28.

West-Eberhardt, M. J. 1979. Sexual selection, social competition, and evolution. *Proceedings of the American Philosophical Society, 51*(4), 222–234.

Westermarck, E. 1891. *The history of human marriage.* London: Macmillan.

Whiten, A., and R. W. Byrne. 1988. *Machiavellian intelligence: Social expertise and the evolution of intellect in monkeys, apes and humans.* New York: Oxford University Press.

Whiten, A., V. Horner, and F. B. M. de Waal. 2005. Conformity to cultural norms of tool use in chimpanzees. *Nature, 437,* 737–740.

Williams, J., H. Liu, and A. Pusey. 2002. Costs and benefits of grouping for female chimpanzees at Gombe. In C. Boesch, G. Hohmann, and L. Marchant, eds., *Behavioural diversity in chimpanzees and bonobos* (pp. 192–203). Cambridge, MA: Cambridge University Press.

Wilson, D. S. 2002. *Darwin's cathedral: Evolution, religion and the nature of society.* Chicago: University of Chicago Press.

Wilson, D. S., and E. O. Wilson. 2008. Evolution "for the good of the group." *American Scientist, 96*(5), 380–389.

Wittig, M., K. Jensen, and M. Tomasello. 2013. Five-year-olds understand fair as equal in a mini-ultimatum game. *Journal of Experimental Child Psychology, 116*(2), 324–337.

Wittig, R. M., C. Crockford, T. Deschner, K. E. Langergraber, T. E. Ziegler, and K. Zuberbühler. 2014. Food sharing is linked to urinary oxytocin levels and bonding in related and unrelated wild chimpanzees. *Proceedings of the Royal Society of London, Series B: Biological Sciences, 281*(1778), 20133096.

Wrangham, R. W., and D. Peterson. 1996. *Demonic males: Apes and the origins of human violence.* Boston: Houghton Mifflin.

Wyman, E., H. Rakoczy, and M. Tomasello. 2009. Normativity and context in young children's pretend play. *Cognitive Development, 24*(2), 146–155.

Yamamoto, S., and M. Tanaka. 2009. Do chimpanzees *(Pan troglodytes)* spontaneously take turns in a reciprocal cooperation task? *Journal of Comparative Psychology, 123*(3), 242–249.

Zahavi, A. 2003. Indirect selection and individual selection in sociobiology: My personal views on theories of social behaviour. *Animal Behaviour, 65*(5), 859–863.

Zeidler, H., E. Herrmann, D. Haun, and M. Tomasello. In press. Taking turns or not? Children's approach to limited resource problems in three different cultures. *Child Development.*

Index

Ache foragers, 46
Agency. *See* Cultural agency; Joint agency, dual-level structure of
Aggression, and self-domestication, 42
Agriculture, 129
Alexander, R. D., 137
Altruism: as unable to evolve, 11; reciprocal, 13–14; in evolution of cooperation, 14–18; and evolution of morality, 137–138
Anger, social, 34
Apes. *See* Great apes
Arbiter, external, 83
Attitudinal reciprocity. *See* Emotional reciprocity

Baumard, N., 138
Biology of Moral Systems, The (Alexander), 137
Bird's eye view on collaborative process, 55–56, 80, 144
Boehm, C., 43, 138
Boesch, C., 27–28, 60–61
Bonobos: sociality and competition among, 21–22; collaboration for competition in, 23–26; collaboration for food among, 26–28; kin- and friend-based prosociality in, 34–35; joint intentionality in, 143–144
Bowles, S., 142
Boyd R., 141–142
Bräuer, J., 33
Brosnan, S. F., 32–33, 166n7
Buttelmann, D., 23

Calculated reciprocity, 14
Callaghan, T., 157
Capitalistic markets, as cooperative cultural institutions, 158
Central-place foraging, 115–116
Chapais, B., 42
Childcare, collaborative, 43

Children: as research subjects, 41; collaborative childcare, 43; collaborative foraging among, 45; helping in, 47–49; joint intentionality in, 51, 52–53; partner choice in, 58; free rider control among, 61; communication in, 65; joint commitments in, 66–67; second-personal protest in, 68–69; resource distribution in, 71, 116–117; guilt and second-personal responsibility in, 74; collaboration in, 76–77; conformity in, 89–90; in-group favoritism among, 91–92; cultural common ground among, 94–95; third-party punishment in, 101; norm enforcement in, 102; normative rules in, 103; obligation to do right in, 108; and creation of moral identity, 111–112; respect property rights, 118; nascent norm-based morality of preschool, 120–121; moral development of contemporary, 154–157; concern of, for others' welfare, 159; joint pretense in, 167n4
Chimpanzees: sociality and competition among, 21–22; collaboration for competition in, 23–26; collaboration for food among, 26–28; sympathy and helping in, 29–31; kin- and friend-based prosociality in, 34–35; as moral agents, 39; partner choice in, 58; free rider control among, 60–61; collaboration in, 65; division of spoils in, 71; joint intentionality in, 143–144; conformity in, 166n1
Chwe, M. S. Y., 93–94
Clutton-Brock, T., 16
Coalitions, among great apes, 23–24, 25
Cognition and cognitive processes: as psychological prerequisite for morality, 35; type I and type II, 140
Collaboration. *See* Cooperation
Collaborative childcare, and self-domestication, 43
Collective commitment, 107, 108–111, 121, 125
Collective common ground, 93–96
Collective intentionality: evolution of, 4–5; and "objective" morality, 92–97; in children, 120; in modern humans, 123, 145; and evolution of morality, 152–153
Common ground, cultural, 93–99
Communication: joint intentionality and emergence of, 53; recognition of cooperative identity through, 63; and joint commitment, 65, 66–67; and second-personal protest, 69–70; in children and great apes, 76. *See also* Language
Competence: partner choice and, 58, 59; and cooperative identity, 63
Competition: cooperation and, 9–10; among great apes, 21–23; collaboration for, in great apes, 23–26; partner choice and, 59
Conformity: culture and, 89–90; identity and, 106; judgment regarding, 108–109; of modern humans, 121; enforcement of, 127; in children, 156; in chimpanzees, 166n1
Contemporary humans: cultural identity of, 90; cultural morality in, 127–128; actions of hypothetically invisible, 160
Contest competition, 21, 26, 34
Conventional cultural practices, 85–86, 96–97
Cooperation: forms of, 1; evolution of human, 3–4, 18–20, 40–41, 135–136; in contemporary human beings, 6–7; competition and, 9–10; evolutionarily stable patterns of, 10–14; interdependence and altruism in evolution of, 14–18; among great apes, 20–34; and kin- and friend-based prosociality, 34–35; and evolution of morality, 39–40; in early humans, 42–50; partner choice and, 57–58; and joint commitment, 64–65; second-personal protest in, 68–69; adaptation for, 76–77; in cultural groups, 85–86; as right, 126; religion and, 131–132; cultural group selection in evolution of,

142; joint intentionality and, 143–144; in social insects, 165n1
Cooperative breeding, 43
Cooperative identity, creation of, 62–63, 80–81
Cooperative rationality, 79, 81–82
Cosmides, L., 137
Coventionalization, 93–96
Crockford, C., 27–28, 31
Cultural agency, 97–107, 119
Cultural common ground, 93–99
Cultural groups: in evolution of morality, 4–5; norm enforcement in, 6; as collaborative enterprise, 85–86; similarity and group identity in, 88–90; in-group favoritism and loyalty in, 90–92; challenges of larger, 98; interdependence in, 145. *See also* Cultural group selection
Cultural group selection: operation level of, 11; in evolution of cooperation, 12, 20, 142; and moral self-governance, 118–119; in modern humans, 129; and changing social norms, 133–134; and cooperation in social insects, 165n1
Cultural identity, 90, 105–107, 136
Cultural institutions, 103–105, 145–146
Cultural morality, 126–128, 129
Cultural norms. *See* Social norms
Culture: loyalty and, 88–92; creation of, 127–128; and evolution of morality, 141–142

Darwall, Stephen, 39, 83
Darwin, Charles, 50
Dawkins, R., 11
Decision making, moral, 113–115, 160–161
Dependency conceptualization, 147
Deservingness. *See* Minimum deservingness; Mutual deservingness
Developmental psychology, 139
Deviance, 6
De Waal, F. B. M.: on reciprocity, 14; findings on reciprocity among great apes, 24–25; on morality in animals, 36; on evolution of morality, 137; findings on fairness in capuchin monkeys, 166n7
Disgust, 133
Distributive justice, and moral self-governance, 115–118
Division of labor, 90–91
Dominance contests, among great apes, 23–24
"Dunbar's number," 88–89
Durkheim, E., 105, 132
Dyadic interactions, 41, 62, 129, 135–136

Early humans: cooperation and helping in, 42–50; partner choice in, 58–59; free rider control among, 61–62; cooperative identity creation among, 63; communication in, 65; second-personal protest in, 68; second-personal responsibility in, 73, 74–75; adapted for collaboration, 76–77; moral psychology of, 78–83, 84; joint intentionality in, 144
Emotion: recognized by great apes, 22–23; in moral decision making, 139–140
Emotional reciprocity, 14, 17, 24–25
Empathy, 49, 137. *See also* Sympathy
Equality, 162
Equivalence. *See* Self-other equivalence
Evolutionary ethics, 137–139
External arbiter, 83

Fairness: morality of, 2, 50, 79–80, 129; in great apes, 32–34; reciprocity as precursor for, 36; as pillar of morality, 36–38; in division of spoils, 68–69, 70–72; and distributive justice, 116–118. *See also* Justice
Favoritism, in-group, 90–92
Food, collaboration for, in great apes, 26–28
Food sharing: cultivation of friends through, 24; as sympathetically motivated, 29; physiological reactions during, 31; social norms governing, 98

Foraging: and self-domestication, 42–43; obligate collaborative, 44–46, 50, 57–58, 81–82, 88, 143–144; partner choice and, 58, 59; central-place, 115–116
Free riders, controlling, 60–62, 80

Gauthier, David, 9
Gene-culture coevolution, 141–143
Gintis, H., 142
Gossip, 100, 138
Gräfenhain, M., 67
Great apes: prosocial behavior among, 3; sociality and competition among, 21–23; collaboration for competition in, 23–26; collaboration for food in, 26–28; sympathy and helping in, 28–31; sense of fairness in, 32–34; kin- and friend-based prosociality in, 34–35; last common ancestor of humans and, 35–36, 38; joint intentionality in, 52, 143–144; partner choice in, 58; adapted for collaboration, 76–77; self-interest in, 159. *See also* Chimpanzees
Greenberg, J. R., 30
Greene, J. D., 141
Grooming, 24, 29, 31
Group augmentation, 16
Group defense, 23
Group identity, 88–90
Group-minded engagement, 136–137, 142–143
Group punishment, 138
Group selection. *See* Cultural group selection
Guilt: and second-personal responsibility, 73–75; and collective commitment, 108–111, 145; conformity and, 127

Haidt, J., 133, 140, 141
Haldane, J. B. S., 11
Hamann, K., 71
Hamilton, W. D., 11
Harm, 139–140
Hawk-dove interaction, 15, 165n3
Helping: in great apes, 28–31; in early humans, 42–50; interdependence and, 47–48, 49, 147–150; in children, 47–49
Henrich, J., 116
Hepach, R., 48
Hopper, L. M., 33
Horner, V., 166n6
House, B. R., 116
Humans: last common ancestor of great apes and, 35–36, 38; moral evolution of, 39–40. *See also* Contemporary humans; Early humans; Modern humans
Hume, David, 37
Hunting: collaboration in, 26–28; as sympathetically motivated, 29; in chimpanzees and bonobos, 34–35; and self-domestication, 42–43; in early humans, 44; interdependence in, 46–47; joint commitment in, 65

Identity: creation of cooperative, 62–63, 80–81; cultural, 90, 105–107, 136; moral, 111–115, 125, 156; and modern human moral psychology, 122
In-group conflicts, 161–162
In-group favoritism, 90–92, 155
Insects, cooperation among, 11, 165n1
Institutions, cultural, 103–104, 145–146
Instrumental helping, in chimpanzees, 29–31
Intentionality. *See* Collective intentionality; Joint intentionality; Shared intentionality
Intentional pedagogy, 96–97
Interdependence: in evolution of cooperation, 14–18, 40; among great apes, 24–25; in hunting, 46–47; helping and, 47–48, 49; and emergence of moral psychology, 78–79; in cultural groups, 85, 90–91, 145; and evolutionary origins of human moral psychology, 147–150; recognition of, 159–160

Interdependence hypothesis, 3–4, 146–147, 153, *153*
Intuition, 139–140

Jensen, K., 30, 32, 36
Joint agency, dual-level structure of, 50–53
Joint commitment: and second-personal morality, 64–75; and division of spoils, 71–72; and second-personal responsibility, 74; division of, in children and great apes, 77; and we>me formula, 81; in cultural groups, 86; in early humans, 144–145
Joint intentionality: evolution of, 4; and second-personal morality, 50–57; and second-personal responsibility, *75;* in children and great apes, 76; and original "ought," 82; and modern human moral psychology, 123; morality and, 143–154
Joyce, R., 154
Justice: distributive, 115–118; procedural, 117–118; morality of, 129. *See also* Fairness
Justification, 113–114

Kin- and friend-based prosociality, 34–38
Kin selection: operation level of, 11; in evolution of cooperation, 11–12, 19–20; interdependence and, 17
Kitcher, P., 133, 138
Korsgaard, Christine, 40, 85
Kropotkin, P. A., 18

Labor, division of, 90–91
Language: in early humans, 59; and cultural common ground, 94–95; generic normative, 97, 102, 103, 121, 145–146; and norm enforcement, 102; reputation and, 138. *See also* Communication
Laws and legal systems, 130–131
Legitimization, and modern human moral psychology, 123–124
Liebal, K., 94
Loyalty, 88–92, 122

Mameli, M., 109
Marriage, 104
Mating, and self-domestication, 42
Mead, George Herbert, 158
Mechanical solidarity, 90
Melis, A. P.: findings on helping in chimpanzees, 24, 30; on cooperation among great apes, 27; findings on collaborative foraging, 44–45
Mikhail, J., 139–140
Minimum deservingness, 116
Mitani, J. C.: on chimpanzee cooperation and competition, 23; findings on food sharing in chimpanzees, 24; on cooperation among great apes, 26, 28
Modern humans: division of labor in, 90–91; collective intentionality in, 92–93, 145; collaboration in, 93; correct and incorrect ways of performing roles among, 96–97; social norms in, 98, 100–101; equal deservingness among, 105–106; conformity in, 108–109, 121; division of resources among, 115–116; cultural group selection in, 118–119; moral psychology of, 122–126; internalization of collective commitment in, 125; morality of, 129–134; interdependence in, 145; social control in, 145–146; social contracts among, 146
Moral-adaptive behaviors, 149
Moral decisions, 113–115, 160–161
Moral identity: and moral self-governance, 111–115; creation and maintenance of, 125, 156
Morality: forms of, 1–2; evolution of, 4–6, 39–40; affecting contemporary human beings, 6–7; psychological prerequisites for, 35–38; Darwin on development of, 50; natural and cultural, 126–128; theories of evolution of, 137–143; as complex result of many processes, 152–153. *See also* "Objective" morality; Second-personal morality
Morality of fairness, 2, 50, 79–80, 129

Morality of justice, 129
Morality of sympathy, 1–2, 79, 129
Moralization, and modern human moral psychology, 124–126
Moral judgments, 140, 160, 167n5
Moral norms: enforcement of, 6; creation of, 86; violation of, 99; and second-personal morality, 122; and modern human moral psychology, 123–124. *See also* Social norms
Moral psychology: creation of novel, 5, 6; of early humans, 78–83; of modern humans, 122–126; evolution of group-minded, 136; and evolution of morality, 139–141
Moral self-governance, 107–121
Moral-structural dimension of human morality, 151–153
Mueller, M. N., 23, 26, 28
Mutual deservingness: and second-personal agency, 61–62; and recognition of self-other equivalence, 80; in modern humans, 105–106; in early humans, 144; interdependence and, 150–151
Mutualism: operation level of, 11; in evolution of cooperation, 13; interdependence and, 18
Mutual respect, 60, 80, 144

Nagel, T., 56
Natural morality, 126–127
Natural selection, 10–14, 147, 158–159
Nichols, S., 99
Normative trust, 64, 66
Norms. *See* Moral norms; Social norms

Objectification, and modern human moral psychology, 123
"Objective" morality, 85–87; evolution of, 6; culture and loyalty and, 88–92; and collective intentionality, 92–97; cultural agency and, 97–107; and moral self-governance, 107–121; in modern humans, 145–146
Obligate collaborative foraging, 44–46, 50, 57–58, 81–82, 88, 143–144
Obligation: Korsgaard on, 85; in children, 121; self-regulation and, 125
Ontogeny, 154–157
Organic solidarity, 90
"Ought," original, 78–83
Out-group conflicts, 161–162

Pair bonding, and self-domestication, 42
Partner choice: in evolution of cooperation, 18–19, 40; in great ape coalition formation, 25; among great apes in hunting, 27–28; and collaborative foraging, 45; self-other equivalence and, 56; and second-personal agency, 57–60; and cooperative identity, 62–63; and evolution of morality, 138; in early humans, 144
Partner control: in evolution of cooperation, 18–19; and second-personal agency, 60–62; in children and great apes, 77
Paternalistic helping, 48, 49
Person, defined, 106
Perspective, cultural common ground and agent-independent, 95–96
Piaget, Jean, 135, 156
Private property, 118
Procedural justice, 117–118
Property, private, 118
Prosociality, kin- and friend-based, 34–38
Proximity, and kin selection, 11

Rakoczy, H., 167n4
Rand, D., 140
Reciprocity: operation level of, 11; in evolution of cooperation, 13–14; calculated, 14; emotional, 14; interdependence and, 18; among great apes, 24–25; as precursor for fairness, 36; versus collaboration, 46–47; and evolution of morality, 137, 138–139
Reflective endorsement, 110, 111

Religion, and modern human morality, 131–133
Reparation, 110
Representative authority, 68, 102, 111
Reputation, 100–101, 138
Resentful protest. *See* Second-personal protest
Resentment: as reaction to unfairness, 33–34; in second-personal protest, 67–70; and division of spoils, 72; for violations of equality and respect, 112–113
Respect, mutual, 60, 80, 144
Richerson, P., 141–142
Right: collective commitment to do, 108, *109;* original, 121–128; cooperation as, 126; moral judgments of, 151–152
Roberts, G., stakeholder model of, 15–17
Rochat, P., 116
Role ideals, 54–55, 150–151
Role interchangeability, 52, 55–56, 62
Role standards, evolution and impartiality of, 4
Roughley, N., 33, 49
Rousseau, Jean Jacques, 1, 37, 64

Sacralization of institutions, 105
Sanctioning, in joint commitments, 67–68
Scavenging. *See* Foraging
Schäfer, M., 116–117
Scramble competition, 21
Searle, J., 104
Second-personal address, 63, 66, 68, 70, 144
Second-personal agency, 40–41, 57–63, 82
Second-personal engagement, 135–137
Second-personal morality: emergence of, 41; collaboration and helping and, 42–50; joint intentionality and, 50–57; second-personal agency and, 57–63; joint commitment and, 64–75; and original "ought," 78–83; possibility of purely, 83–84; and transition to "objective" morality, 86; and distributive justice, 117; and moral norms, 122; in modern humans, 129; natural, 160
Second-personal protest, 67–70
Second-personal responsibility: guilt and, 73–75; in early humans, 145
Self-control, exercised by great apes, 22
Self-domestication, of early humans, 42–43
Self-governance, moral, 107–121
Self-interest, 158–159
Self-other equivalence: and second-personal morality, 55–57; and mutual respect, 60; recognition of, 80, 162; in modern humans, 105–106; interdependence and, 151–152
Self-regulation: as psychological prerequisite for morality, 35; in modern humans, 125, 146; in early humans, 144–145; interdependence and, 151–152
Sex, social norms governing, 98–99
Sexual selection, 18–19
Shame, 110
Shapiro, S., 130–131
Shared intentionality, 4, 13, 50, 136, 150, *153*
Shweder, R. A., 141
Silk, J. B., 30, 36
Similarity, group identity and, 88–90
Smith, Adam, 69
Smith, Maynard, 15
Smithian empathy, 49, 79
Sober, E., 137–138
Social anger, 34
Social contract, 78, 106–107, 146
Social imitation, 89–90
Social motivation, 35
Social norms: emergence and enforcement of, 6, 98–103; as moral, 100, 126–127; violation of, 112–113; and distributive justice, 117; and cultural group selection, 118; in children, 120–121, 155–156; and modern human moral psychology, 123–124; changing, 133–134; of modern humans, 145–146. *See also* Moral norms

Social planning, 130–131
Social psychology, 139
Social selection: in evolution of cooperation, 18–19; in great ape coalition formation, 25; and evolution of morality, 138
Spoils, division of, 68–69, 70–72, 76, 77
Stakeholder model, 15–17
Status functions, 104–105
Strategic-adaptive behaviors, 149
Strawson, P. F., 83
Supraindividual entity, 64, 81, 108
Supraindividual social structures, 78
Symbiosis, 17
Sympathy: morality of, 1–2, 79, 129; and kin selection, 11–12; and interdependence among great apes, 25; in great apes, 28–31; as precursor for fairness, 36; origination of, 45–46; and helping in children, 49. *See also* Empathy

Taboo, 105
Tanaka, M., 30
Third-party punishment, 101–102
Tomasello, M., 29, 31, 167n4
Tooby, J., 137
Tribal organization, emergence of, 88
Trust: in formation of joint goal, 51; normative, 64, 66
Tuomela, R., 27
Type I cognitive processes, 140
Type II cognitive processes, 140

Ultimatum game, 32, 33

"Voice," 133–134

War, 161–162
Warneken, F.: findings on helping in chimpanzees, 29, 30, 31; findings on collaborative foraging, 45; joint commitment study of, 66–67
Watts, D., 24
We>me formula, 81–82, 145, 146, 151
Williams, J., 25–26
Wilson, D. S., 132, 137–138
Wittig, R. M., 31
Wrong: normative rules of, 85–87, 96–97, 102–103; collective commitment to right and, 108, *109;* original, 121–128; moral judgments of, 152
Wyman, E., 167n4

Yamamoto, S., 30
You=me formula, 79–81
You>me formula, 79

Zahavi, A., 17
Zeidler, H., 116